Monographs on Theoretical and Applied Genetics 12

Edited by

R. Frankel (Coordinating Editor), Bet-Dagan

M. Grossman, Urbana · H. F. Linskens, Nijmegen

P. Maliga, Oakland · R. Riley, London

Monographs on Theoretical and Applied Genetics

Volume 1 **Meiotic Configurations**
A Source of Information for Estimating Genetic Parameters
By J. Sybenga (1975)

Volume 2 **Pollination Mechanisms, Reproduction and Plant Breeding**
By R. Frankel and E. Galun (1977)

Volume 3 **Incompatibility in Angiosperms**
By D. de Nettancourt (1977)

Volume 4 **Gene Interactions in Development**
By L. I. Korochkin (1981)

Volume 5 **The Molecular Theory of Radiation Biology**
By K. H. Chadwick and H. P. Leenhouts (1981)

Volume 6 **Heterosis**
Reappraisal of Theory and Practice
Editor: R. Frankel (1983)

Volume 7 **Induced Mutations in Plant Breeding**
By W. Gottschalk and G. Wolff (1983)

Volume 8 **Protoplast Fusion**
Genetic Engineering in Higher Plants
By Y. Y. Gleba and K. M. Sytnik (1984)

Volume 9 **Petunia**
Editor: K. C. Sink (1984)

Volume 10 **Male Sterility in Higher Plants**
By M. L. H. Kaul (1988)

Volume 11 **Tree Breeding: Principles and Strategies**
By G. Namkoong, H. C. Kang, and J. S. Brouard (1988)

Volume 12 **The Wheat Rusts - Breeding for Resistance**
By D. R. Knott (1989)

Douglas R. Knott

The Wheat Rusts — Breeding for Resistance

With 19 Figures and 40 Tables

Springer-Verlag
Berlin Heidelberg New York
London Paris Tokyo

Professor Dr. DOUGLAS R. KNOTT
College of Agriculture
University of Saskatchewan
Saskatoon, Saskatchewan, Canada S7N OWO

ISBN-13:978-3-642-83643-5 e-ISBN-13:978-3-642-83641-1
DOI: 10.1007/978-3-642-83641-1

Library of Congress Cataloging-in-Publication Data. Knott, Douglas R., 1927 – The wheat rusts: breeding for resistance/Douglas R. Knott. – (Monographs on theoretical and applied genetics; 12)
ISBN-13:978-3-642-83643-5
1. Wheat – Disease and pest resistance – Genetic aspects. 2. Wheat rusts. 3. Wheat – Breeding. I. Title. II. Series. SB608.W5K58 1989. 633.1'19425-dc19 88-8075

Softcover reprint of the hardcover 1st edition 1989

Typesetting: International Typesetters, Inc., Makati, Philippines
Offsetprinting: Color-Druck Dorfi GmbH, Berlin
Bookbinding: Bruno Helm, Berlin
2131/3020/543210 – Printed on acid-free paper

This book is dedicated to the late

Dr. Irvine A. Watson,

a very enthusiastic and productive wheat scientist,
who made many important contributions
in the field of rust research, and who conceived
the idea for this book.

Legend for Frontispiece

Uredia of leaf rust (**a** and **b**), stem rust (**c** and **d**), and yellow rust (**e** and **f**) on wheat plants; photographs **a** to **d** courtesy of Dr. P. L. Dyck, Agriculture Canada Research Station, Winnipeg, and **e** and **f** courtesy of AFRC-IPSR Cambridge Laboratory

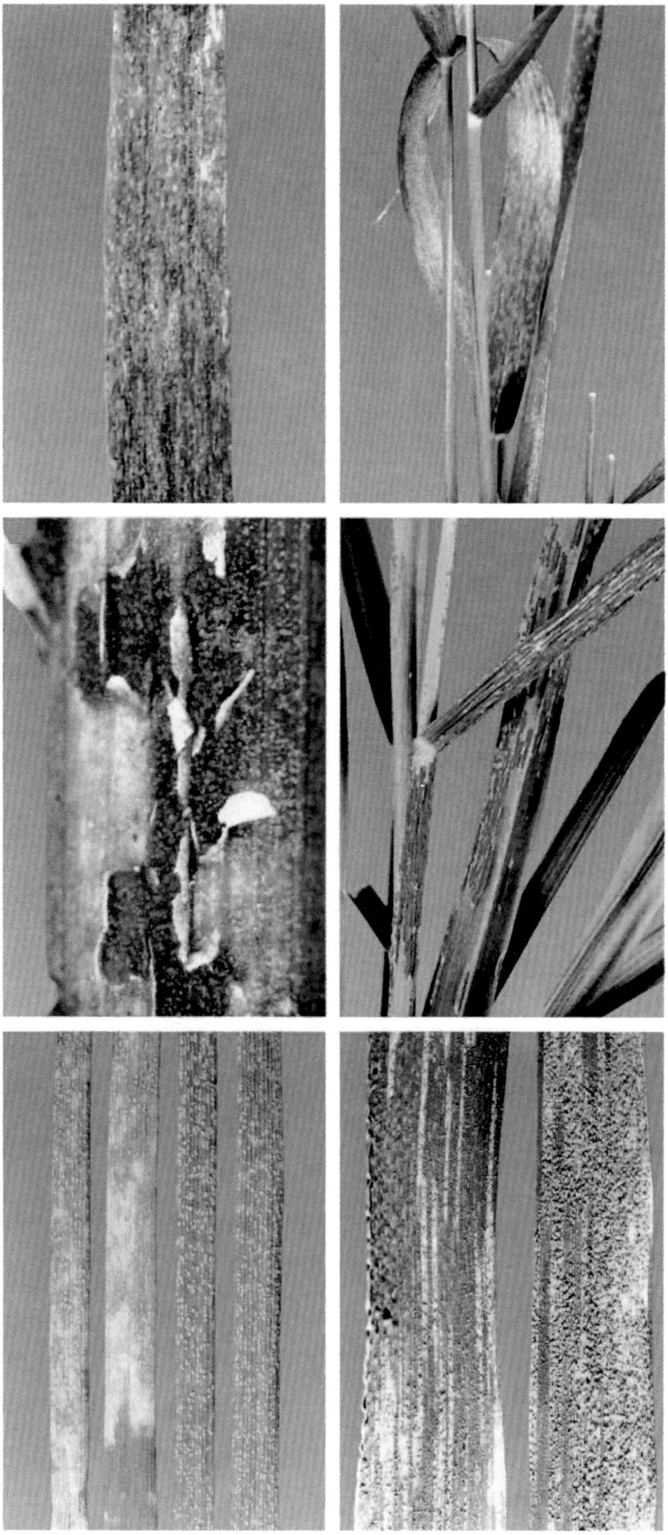

a b c d e f

Preface

The idea for this book was conceived by the late Dr. Irvine A. Watson of the University of Sydney, and he developed the first outline. I was then invited by Dr. Watson to share in its writing. Unfortunately, shortly thereafter, recurring heart problems forced him to curtail his activities and withdraw from the project. He died before the book could be completed.

Dr. Watson's intention was to produce a very practical book that would provide wheat breeders with all of the information necessary to breed successfully for resistance to the three wheat rusts: leaf rust, stem rust, and yellow rust. It was intended to be very specific in describing procedures to be used and at the same time provide all of the necessary theoretical background. I hope that I have been successful in meeting these objectives. The book assumes that the reader has some knowledge of plant pathology, genetics, and plant breeding.

Extensive use has been made of the literature, but it was not possible to cite all of the papers on a given topic. In making a choice, an attempt was made to choose key papers or more recent papers that provided references to the earlier literature.

Acknowledgements

This book was written partly at the University of Saskatchewan and partly at the University of California, Davis, while I was on a sabbatical leave. I am particularly grateful to the Department of Agronomy and Range Science and its chairman, Dr. D. A. Phillips, for providing me with excellent facilities and support during my stay at Davis.

I am very grateful to friends and colleagues who reviewed chapters for me: Dr. P. L. Dyck, Dr. P. McGuire, Dr. R. W. Michelmore, Dr. C. O. Qualset, and especially Dr. R. Johnson, who read several chapters and collaborated on the section on the genetics of resistance to yellow rust. Nevertheless, I must take responsibility for any errors that may occur and for the opinions that are expressed.

I very much appreciated the help of typists who translated my scrawl into print: Betty Ann Atimoyoo, Mary Lee, and Joan McLean at the University of Saskatchewan, and Linda Davis, Marilee Schmidt, and Evelyn Martinelli at the University of California, Davis.

Saskatchewan, Canada
January 1989

DOUGLAS R. KNOTT

Contents

CHAPTER 1
Introduction

1.1 The Hosts

The term wheat is normally used to refer to the cultivated species of the genus *Triticum*. The genus *Triticum* is complex and includes diploids, tetraploids, and hexaploids. Although a number of species have been cultivated over the years, cultivation is now restricted almost entirely to the tetraploid durum wheat (*Triticum turgidum* L.) and the hexaploid common or bread wheat (*Triticum aestivum* L.).

1.1.1 The Origin and Evolution of Wheat

Sakamura (1918) showed that the wheats fall into three groups with chromosome numbers of 14, 28, and 42. Thus, the basic chromosome number and the number of chromosomes in a genome is 7. Following Sakamura's work, Kihara (1919, 1924) began his very extensive investigation of wheat and its relatives, using genome analysis (reviewed by Lilienfeld 1951). Sax (1922) also studied the cytological behavior of crosses between the three chromosome groups.

In genome analysis, two species are crossed and the pairing of their chromosomes at meiosis is used to determine whether they carry the same or different genomes. Diploids with known genomes can be used to identify genomes in other diploids or polyploids. For example, when the diploid einkorn wheat (*Triticum monococcum* L.) is crossed with the tetraploid durum wheat, the meiotic configuration is often $7'' + 7'$ (7 pairs plus 7 univalents). Therefore, durum wheat carries one genome (A) in common with einkorn wheat and has one additional genome (B). A cross between durum and bread wheats often has $14'' + 7'$ at meiosis. Thus, bread wheat carries both the A and B genomes plus a third genome (D). The pairing shown is the maximum.

The A genomes present in the three species have been separated for many generations. Differentiation, which may include translocations, has undoubtedly occurred. In a cross between einkorn and durum wheat, fewer than seven pairs may occur or there may be a quadrivalent as a result of a translocation. After long periods of evolution, common genomes in different species can differ to varying degrees. Nevertheless, genome analysis was very effective in establishing the relationships among the genomes in the *Triticum* species and their relatives. More recently, sophisticated mathematical methods of genome analysis have been developed (e.g., Driscoll et al. 1980, Kimber 1983).

The genus *Triticum* provides a textbook example of the evolution of species through amphiploidy. Based on genome analysis, it was evident that the cultivated *Triticums* evolved as shown in Fig. 1.1. The donor of the D genome was determined

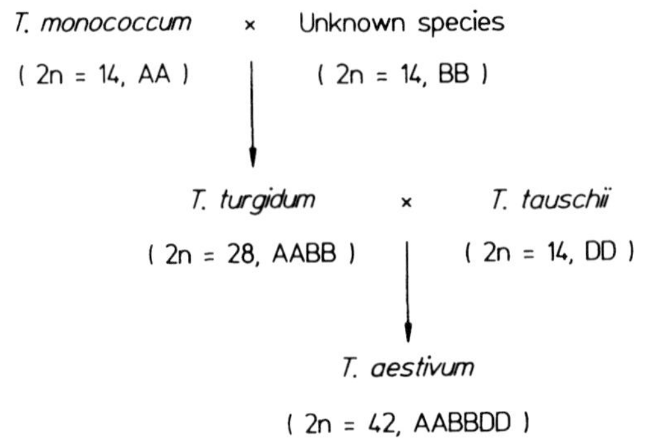

Fig. 1.1. The evolutionary origin of *Triticum turgidum* (durum wheat) and *Triticum aestivum* (bread wheat)

to be *Aegilops squarrosa* (now often classified as *T. tauschii*) (reviewed by Riley 1965). However, the donor of the B genome has not been identified, although a species in the Sitopsis section of the genus *Aegilops* (now often included in *Triticum*) has been suggested (reviewed by Miller 1987 and Kerby and Kuspira 1987). It may be that the donor is extinct and only species that have evolved from it exist today.

1.1.2 The Classification of Wheat

Originally the genus *Triticum* included only those species that had the A genome in common. Each of the three ploidy levels was divided into two or more species based primarily on morphological characters, as follows:

A. Diploids (2n = 14, AA) – einkorn wheats
- *T. boeoticum* Boiss. – wild einkorn wheat
- *T. monococcum* L. –einkorn wheat

B. Tetraploids (2n = 28, AABB) – durum and emmer wheats
- *T. durum* Desf. – durum wheat
- *T. dicoccum* Schrank – emmer wheat
- *T. dicoccoides* Korn. – wild emmer wheat
- *T. turgidum* L. – poulard, rivet or cone wheat
- *T. polonicum* L. – Polish wheat
- *T. carthlicum* Nevski (*T. persicum* Vav.) – Persian wheat

T. timopheevii is also a tetraploid, but is now generally considered to have the genomic formula AAGG.

C. Hexaploids (2n = 42, AABBBDD) – common or bread wheats
- *T. aestivum* L. – common or bread wheats
- *T. compactum* Host. – club wheat
- *T. spelta* L. and *T. macha* Dek. & Men. – spelt wheat
- *T. sphaerococcum* Perc. – shot wheat
- *T. vavilovii* Jakubz.

T. zhukovskyi Men. & Er. is also a hexaploid but has the genomic formula AAAAGG.

Most species in all three groups have been cultivated to some extent, although only *T. durum* and *T. aestivum* are cultivated extensively now. Since the theme of this book is breeding for rust resistance, it will deal almost exclusively with durum and bread wheat, except when other species are used as sources of resistance.

The discovery that two of the wheat genomes came from *Aegilops* species made a revision of the *Triticum* genus necessary. Although there is still not complete agreement among taxonomists, many now include in *Triticum* all of the species that were formerly classified as *Aegilops*. The classification used by Kimber and Sears (1987) is reproduced in Tables 1.1 and 1.2. In addition to the placing of the *Aegilops* species in *Triticum*, a second major change is the consolidation into single species of all of the former *Triticum* species having the same ploidy level. The exceptions are *T. timopheevii* and *T. zhukovskyi*, which are maintained as separate species because they carry the G genome rather than B. Within a ploidy level, all of the original *Triticum* species cross readily and produce fertile hybrids.

1.1.3 The Homoeologous Chromosomes in Bread Wheat

The proposed origin of bread wheat suggests that it has three closely related genomes and, therefore, each chromosome in one genome should be related or homoeologous to one in each of the other two genomes. This is exactly what Sears (1952, 1954) found when he tested the ability of a tetrasomic for one chromosome to compensate for a

Table 1.1. The proposed genome symbols of the diploid species of *Triticum*

Species	Symbol	Synonyms
T. monococcum L.	A	*T. boeoticum* *T. urartu*
T. speltoides (Tausch) Gren. ex Richter	S	*Ae. speltoides*
T. bicorne Forsk.	S^b	*Ae. bicornis*
T. longissimum (Schweinf. & Muschli in Muschli) Bowden	S^l	*Ae. longissima*, *Ae. sharonensis*
T. searsii (Feldman & Kislev) Feldman, comb. nov.	S^s	*Ae. searsii*
T. tripsacoides (Jaub. & Spach) Bowden	Mt	*Ae. mutica*
T. tauschii (Coss.) Schmal.	D	*Ae. squarrosa*
T. comosum (Sibth. & Sm.) Richter	M	*Ae. comosa*, *Ae. heldrechii*
T. uniaristatum (Vis.) Richter	Un	*Ae. uniaristata*
T. dichasians (Zhuk.) Bowden	C	*Ae. caudata*
T. umbellulatum (Zhuk.) Bowden	U	*Ae. umbellulata*

Table 1.2 The proposed genome symbols of the polyploid species of the genus *Triticum*. Italized symbols indicate substantial modifications of the genomes concerned

Species	Genomes	Synonyms
T. turgidum L.	AB	*T. carthlicum, T. dicoccoides, T. dicoccon, T. dicoccum T. durum, T. polonicum*
T. timopheevii (Zhuk.) Zhuk.	AG	*T. araraticum*
T. zhukovskyi Men. & Er.	AAG	*T. timopheevii* var. *zhukovskyi*
T. aestivum L.	ABD	*T. compactum, T. macha, T. spelta, T. sphaerococcum, T. vavilovii*
T. ventricosum Ces.	DUn	*Ae. ventricosa*
T. crassum(4x) (Boiss.) Aitch. & Hensl.	D*M*	*Ae. crassa*
T. crassum(6x) (Boiss.) Aitch. & Hensl.	*DDM*	*Ae. crassa*
T. syriacum Bowden	*DMS*	*Ae. crassa* ssp. *vavilovii, Ae. vavilovii*
T. juvenale Thell.	*DMU*	*Ae. juvenalis*
T. kotschyi (Boiss.) Bowden	U*S*	*Ae. kotschyi, Ae. peregrina, Ae. variabilis*
T. ovatum (L.) Raspail	U*M*	*Ae. ovata*
T. triaristatum(4x) (Willd.) Godr. & Gren.	U*M*	*Ae. triaristata*
T. triaristatum(6x) (Willd.) Godr. & Gren.	U*MUn*	*Ae. triaristata*
T. machrochaetum (Schuttl. & Huet. ex Duval-Jouve) Richter	U*M*	*Ae. biuncialis, Ae, lorentii*
T. columnare (Zhuk.) Morris & Sears	U*M*	*Ae. columnaris*
T. triunciale (L.) Raspail	UC	*Ae. triuncialis*
T. cylindricum Ces.	CD	*Ae. cylindrica*

nullisomic for another in terms of plant vigor and fertility (a tetrasomic has four copies of one chromosome, a nullisomic has none). The chromosomes, which originally had been designated I – XXI, could be placed in seven homoeologous groups of three chromosomes as follows:

1 – I, XIV, XVII   5 – V, IX, XVIII
2 – II, XIII, XX   6 – VI, X, XIX
3 – III, XII, XVI   7 – VII, XI, XXI
4 – IV, VIII, XV

Within a group, the tetrasome for one chromosome compensated for the nullisome of either of the other two chromosomes. No compensation occurred when a tetrasome for a chromosome in one group was combined with a nullisome for a chromosome in a different group.

Because bread wheat carries three homoeologous genomes, many of its genetic systems are triplicated. As a result, a gene in one chromosome may be masked by genes in the two homoeologous chromosomes. For example, none of the nullisomics is deficient in chlorophyll, although one should be deficient if a single locus is involved (Sears 1954). Many genes should have dosage effects and this is what Sears (1966b) discovered when he tested many compensating nullisome-tetrasomes.

1.1.4 Identification of the Chromosomes in the Three Wheat Genomes

The identity of the D genome chromosomes was easily determined by crossing each of the 21 monosomic lines developed by Sears (1954) with a tetraploid wheat (Sears 1959). If the 34-chromosome F_1 plants from a cross showed 14″ + 6′ at metaphase of the first meiotic division, then the monosomic chromosome in the parent belonged to the D genome, i.e., there were 14″ of A and B genome chromosomes and 6 D chromosome univalents. If the 34-chromosome F_1 plants showed 13″ + 8′, then the monosomic chromosome belonged to either the A or B genome, i.e., there were 7 D chromosome univalents plus 1 A or B chromosome univalent. Originally the A and B genome chromosomes were designated I-XIV and the D genome chromosomes XV-XXI.

The identification of the A and B genome chromosomes was more difficult. However, the development of ditelosomic line in which a particular pair of chromosomes had been replaced by a pair of telocentric chromosomes (Sears 1966a), made it possible (a telocentric chromosome is composed of one chromosome arm). Okamoto (1962) crossed 13 ditelosomic lines representing all of the A and B genome chromosomes except IV, with an AADD amphiploid, *T. aegilopoides* (= *T. monococcum*)/*Ae. squarrosa* (= *T. tauschii*). If the telosome was from an A genome chromosome, the F_1 plants showed a heteromorphic bivalent (the telosome paired with a normal A chromosome). If the telosome was from a B chromosome, it did not pair and was present as a telocentric univalent in the F_1 plants. Okamoto (1962) was able to identify 12 of the A and B genome chromosomes in this way. The results for chromosomes II and XIII were ambiguous, although he tentatively designated them as 2B and 2A. However, Chapman and Riley (1966) crossed ditelocentric lines for chromosomes I-XIV with *T. thaoudar* (= *T. monococcum*) (AA) and determined whether the telocentrics did or did not pair with an A genome chromosome. They confirmed all of Okamoto's designations except that II proved to be 2A and XIII was 2B. The chromosomes were then renamed as shown in Table 1.3. However, there is still some controversy about the designation of 4A and 4B, since neither chromosome pairs with a chromosome of *T. monococcum* (Kimber and Sears 1987).

1.1.5 The Control of Homoeologous Pairing in *Triticum*

The various genomes in the genus *Triticum* are assumed to have evolved from a single genome in a common ancestor. While they no doubt differ greatly in the degree to which they are differentiated from one another, the absence of pairing between homoeologous chromosomes in tetraploids and hexaploids is surprising.

Table 1.3. The homoeologous chromosome groups in *Triticum aestivum*

Homoeologous group	Chromosome designations		
	Genome A	Genome B	Genome D
1	1A (XIV)	1B (I)	1D (XVII)
2	2A (II)	2B (XIII)	2D (XX)
3	3A (XII)	3B (III)	3D (XVI)
4	4A (IV)	4B (VIII)	4D (XV)
5	5A (IX)	5B (V)	5D (XVIII)
6	6A (VI)	6B (X)	6D (XIX)
7	7A (XI)	7B (VII)	7D (XXI)

Even in a haploid when no homologous chromosomes are present, pairing among homoeologues is infrequent. The explanation for this was discovered almost simultaneously by Okamoto (1957), Sears and Okamoto (1958) and Riley and Chapman (1958). Sears and Okamoto (1958) crossed Chinese Spring monotelo V (20″+t′5B) X AADD (a *T. aegilopoides* X *Ae. squarrosa* amphidiploid). The 34-chromosome F_1 plants (lacking chromosome V=5B) showed many fewer univalents (8.24) than the 35-chromosome F_1 plants (23.82), and also had many more bivalents and multivalents. Riley and Chapman (1958) found that in the cultivar Holdfast nullihaploid plants (20 chromosomes, lacking 5B) showed much higher pairing, including multivalents, than did haploid plants (21 chromosomes). Both sets of results indicated that in the absence of chromosome 5B, pairing among homoeologous chromosomes occurred. A number of other promoters and suppressors of pairing have been identified, but the major effect is due to a gene on the long arm of 5B (5BL) which has been designated *Phl* (see Kimber and Sears 1987; and Sears 1976, 1984).

1.1.6 Variability in Durum and Bread Wheats

The cultivated durum and bread wheats are grown over wide areas in many different environments. Both species are very diverse and many thousands of cultivars of each are known. Tremendous variability exists for resistance to all three wheat rusts, stem rust (*Puccinia graminis* Pers. f. sp. *tritici* Eriks. and Henn.), leaf rust (*Puccinia recondita* Rob. ex Desm. f. sp. *tritici*) and yellow rust (*Puccinia striiformis* West.).

1.2 The Pathogens

Several thousands of rust species attack a wide range of higher plants. A number of them cause serious economic losses in crops, but none more so than the three rusts that attack wheat, the world's most important crop.

All three rusts belong to the genus *Puccinia* but they differ in morphology, life cycle, and optimal environmental conditions for growth.

1.2.1 Wheat Stem Rust

Stem rust, which is also called black rust, is caused by *Puccinia graminis* Pers. The name black rust refers to the black teliospores which are formed towards the end of the growing season. Within *P. graminis*, specialization on particular host genera has occurred to produce formae speciales. Three of the most important are: *P. graminis* f. sp. *avenae*, which is specific to oats and some related grasses, *P. graminis* f. sp. *secalis* on rye and some related grasses, and, most important, *P. graminis* f. sp. *tritici* on wheat, barley, and many of the relatives of wheat. Usually a forma specialis will not produce pustules on a species outside its normal host range. For example, oat stem rust will normally produce at most small flecks on wheat or occassionally very minute pustules.

Stem rust is considered to be the most damaging of the wheat rusts. It can attack all of the above ground parts of the plant, leaves, leaf sheaths, stems and spikes, including even the awns. Damage to plants results from the loss of photosynthetic area, the disruption of water and nutrient transport, reduction in root growth, and lodging and stem breakage. Depending on the earliness and severity of the attack, stem rust can cause various amounts of kernel shrivelling and yield loss. Lodging can make the crop difficult to harvest. Severe, early attacks can cause total loss of a crop. The same environmental conditions that favor crop growth, particularly ample moisture, also favor rust development. Thus, a particularly promising crop can be destroyed by stem rust in a matter of weeks.

On wheat stems or leaf sheaths, stem rust produces dark brownish-red, elongate pustules. The pustules burst through the epidermis and remnants of it give the pustules a ragged appearance. On leaves, the pustules can be of various sizes and shapes but on young leaves of fully susceptible plants they are often diamond-shaped. On older leaves the pustules tend to be restricted by the veins. Pustules sporulate on both leaf surfaces but tend to be heavier on the lower surface. Secondary spread of the fungus can produce a secondary ring of sporulation outside a primary leaf pustule.

1.2.2 Wheat Leaf Rust

Wheat leaf rust, sometimes called brown rust, is caused by *Puccinia recondita* Rob. ex Desm. It also shows specialization for specific hosts and the wheat leaf rust fungus is commonly designated *P. recondita* f. sp. *tritici*.

Leaf rust is probably the commonest and most widely distributed of the wheat rusts. It primarily attacks the leaf blades and to a lesser extent leaf sheaths and glumes. The primary damage results from premature defoliation of the plants which results in the shrivelling of kernels. Although total crop loss does not occur, yield reductions of up to 40% have been reported. Leaf rust tends to cause less severe damage than stem rust, but in some areas it occurs more frequently and overall can cause greater losses.

The typical symptoms of leaf rust are small, round, orange-red pustules, often about 0.2 cm in diameter. The pustules are largely on the upper leaf surface. They are readily distinguishable from stem rust pustules on leaves by their smaller size, round shape, and orange-red color. The surface layer of spores may darken but it can be wiped off with a finger to reveal the true color. In a severe epidemic, almost the entire surface of the leaf blades can be covered with pustules. The leaves senesce rapidly and dry out, depriving the plant of much of its photosynthetic area.

1.2.3 Wheat Yellow Rust

Yellow rust, which is often called stripe rust, is caused by *Puccinia striiformis* West. In contrast to stem and leaf rust, yellow rust requires relatively cool temperatures for good growth. As a result it is most common in regions where the climate is temperate due to high latitudes or elevation, or in tropical areas where wheat is grown in the cool, moist winter season. For example, in the central plains of North America, yellow rust is often found in the spring, particularly in the south near the mountains. However, temperatures rise rapidly later in the spring and little further spread occurs. However, in the Pacific coast states of the U.S.A., particularly Washington and Oregon which are cooler during the growing season, yellow rust can be a severe problem.

Both names for the rust are descriptive. Typical symptoms are long, yellow stripes on the leaves. All parts of the plant can be attacked, even kernels. On older leaves the pustules are restricted by veins but may grow several inches in length. On seedling leaves, lateral spread of the pustules is less restricted.

As with leaf rust, the primary loss results from defoliation and shrivelling of the kernels. Losses of up to 75% have been reported.

1.3 The Importance of the Rusts

All three rusts are present essentially everywhere wheat is grown. Their importance from area to area depends on climate and also on the degree of resistance of the predominant cultivars. Within an area, variation from year to year depends on the weather. The optimum environments for the three rusts are sufficiently different that one may thrive where the others do not. The biggest difference is the lower temperature optimum and maximum for yellow rust compared to the other two.

The wheat-growing areas of the world can be divided into epidemiological zones within which relatively free movement of the rust spores occurs (Saari and Prescott 1985). The virulence of a rust population will tend to be similar throughout a zone. However, changes can occur from area to area within a zone if the predominant cultivars in different areas carry different genes for resistance. Spores can also be exchanged between some of the zones, with the frequency depending on the degree of geographic isolation.

1.3.1. North America

Stem rust seems to have been a problem in the eastern U.S.A almost as soon as wheat production began. As wheat production moved westward and large acreages were grown on the central plains of the U.S.A. and Canada, stem rust epidemics became frequent. Eventually they were largely controlled by the development and production of resistant cultivars. The last major epidemic occurred in 1954, when both leaf and stem rust were severe (Roelfs 1978). Nevertheless, all three rusts are present every year.

In North America, high temperatures which favor the development of stem rust occur frequently during the reproductive phase of wheat growth. As long as free moisture is present for at least part of the day or night, spore germination and infection occurs. The rust spreads rapidly and can cause severe losses. In 1954 a combined epidemic of leaf and stem rust probably caused losses of more than $500,000,000 in Canada and the U.S.A. For 1973 to 1977, Green and Campbell (1979) estimated the annual saving from growing rust-resistant cultivars on the Canadian prairies to be $217 million. Stem rust can also be important in Mexico but has been controlled by resistant cultivars in recent years.

The spectacular losses caused by stem rust in some years have tended to overshadow the importance of leaf rust. However, because it frequently overwinters over larger areas, leaf rust is often more widespread than stem rust. Damage is less severe but occurs more often than from stem rust. Losses of 5 or 10% are easily overlooked in years when conditions are favorable for both leaf rust and good wheat yields. Leaf and stem rust tend to have fairly similar environmental requirements. In years such as 1954 when both were severe, it is almost impossible to separate the effects of the two.

The potential for severe epidemics of stem and leaf rust still exists in North America, as was shown in 1986. Both rusts built up early in the central plains of the U.S.A., although heavy damage did not occur. Enormous spore loads were blown into the spring wheat areas of the north-central U.S.A. and Canada. Only the presence of resistant cultivars prevented enormous losses. Even then, a large percentage of the leaf area of resistant cultivars was covered with yellow flecks from leaf rust infections. The reduction in photosynthesis undoubtedly caused some yield loss from shrivelling of kernels. Late winter wheat crops in the area were severely damaged, primarily by stem rust. Leaf rust can also be important in Mexico. A severe epidemic occurred in irrigated wheat in the northwestern Mexican states in 1977.

In the U.S.A., yellow rust is important only on fall-planted wheat in the western states, primarily Washington, Oregon, Idaho, and northern California. In these areas the wheat either matures before the weather gets hot or the late spring and early summer are cool so that temperatures are favorable. Yellow rust is occasionally found in the southern Great Plains in the spring but its spread is restricted as temperatures rise. Yellow rust is common in Mexico and can cause damage in local areas at higher elevations.

1.3.2 South America

South America is divided into two subzones by the Andes mountains. The western subzone includes Columbia, Ecuador, Peru, Chile, and western Bolivia and is characterized by tremendous variation in elevation and by primitive agriculture in many areas. The southeastern subzone includes the main wheat-growing areas of the pampas of Argentina, Uruguay, Paraguay, southern Brazil, and lowland Bolivia. The two subzones are largely isolated from one another by the Andes but spores can pass from one to the other through southern passes (Saari and Prescott 1985).

All three rusts are present throughout South America and probably are endemic in many areas. Yellow rust is common in the western subzone and in southern Argentina. It can cause serious losses, particularly in local areas at higher elevations. Leaf rust is widespread and can cause significant damage. Breeding for leaf rust resistance has gone on for many years and South American cultivars have been used as sources of resistance in other countries. Stem rust is also widespread but damage is usually not heavy except in local areas. However, a severe epidemic occurred in Brazil in 1976. Breeding for resistance has been important for many years.

1.3.3 Europe, North Africa, and Western Asia

This is a large, heterogeneous area. It can be divided into eastern and western subzones but considerable spore exchange occurs between the two.

1.3.3.1 Western Europe and Northwestern Africa

In this subzone, all three rusts are present but yellow rust is the most important. Overwintering of the rusts probably occurs in all but the most northern part of the zone. Oversummering is a problem in the south but occurs at higher elevations such as the Atlas mountains.

The cool temperatures of much of western Europe favor yellow rust and it has occurred widely in many years. However, the development of resistant cultivars and the use of fungicides has limited damage in recent years.

Leaf rust is also commonly found in western Europe but losses are generally light.

A century or two ago stem rust was an important disease on wheat, often spreading from the alternate host, barberry. However, it is now of minor importance.

1.3.3.2 Eastern Europe, Western Asia, and Egypt

All three rusts apparently survive the year round in at least some areas of this subzone.

Leaf rust is widespread and has caused damage in eastern Europe and the U.S.S.R. and also in Egypt in recent years. Stripe rust is also common and can cause losses when the weather is favorable. Stem rust is now a relatively minor disease, partly because of the elimination of barberry.

1.3.4 Southern Africa

This zone includes the countries south of the Sahara desert plus the south-western part of the Arabian peninsula. Wheat is generally grown in limited areas that are often geographically isolated so that the exchange of spores between areas is restricted. The largest areas of production are in Ethiopia and South Africa. Much of the wheat is grown at elevations above 1000 m and during rainy seasons that provide relatively cool temperatures. At higher elevations, particularly in Ethiopia and Kenya, yellow rust can cause severe losses in local areas. Stem rust is a problem in the same countries, and also in South Africa where yellow rust is not a problem. Leaf rust occurs in many of the countries but is of minor importance.

1.3.5 The Indian Subcontinent

This epidemiological zone includes India, Pakistan, Bangladesh, Nepal, and Afghanistan (Saari and Prescott 1985). All three rusts survive the hot summers at higher elevations in the north and the Nilgiri Hills in southern India. Saari and Prescott (1985) consider leaf rust to be the most serious of the rusts. It occurs frequently in the northern part of the area and can cause substantial losses. Stem rust is more important in the south and has caused severe epidemics in the past. In recent years the growing of resistant cultivars has largely prevented losses except on local landraces. Because of its requirement for cooler temperatures, yellow rust is important in the north and at higher elevations in southern India. Many cultivars are resistant and epidemics are infrequent.

1.3.6 The Far East

China is the major wheat producer in this area, which also includes Japan, Korea, Mongolia, and the eastern U.S.S.R. In China, spring wheats are grown primarily in the northeast and winter wheats in the rest of the country.

Stripe rust is a major problem throughout the winter wheat area. It oversummers at some of the higher elevations in western China. Until the 1960's, stem rust had been the most important rust, starting each year in the south and moving up into the spring wheat area. However, resistant cultivars have reduced the losses. Leaf rust is also widespread but causes limited damage. China is a probable inoculum source for the other countries in the zone, particularly in the north where rusts may not overwinter.

1.3.7 Southeast Asia

This zone includes Burma, Thailand, Malaysia, Indonesia, and the Philippines, countries with limited wheat production. The rusts are relatively unimportant. Leaf rust is found in most areas and stem rust in Burma and Thailand. Despite the proximity of the zone to India and Bangladesh, its rusts seemed to have developed in isolation (Saari and Prescott 1985).

1.3.8 Australia and New Zealand

In Australia wheat is grown in the southwest and in a larger area in the southeast. The two areas are separated by about 1300 km of desert but some spore exchange occurs. Limited amounts of wheat are grown in New Zealand. Although separated from Australia by about 2000 km of the Tasman Sea, spores are regularly received from eastern Australia (Luig 1985). In Australia, spring habit wheats are grown during the mild winter. The rusts apparently oversummer on volunteer wheat and some grasses.

When weather conditions are favorable, severe stem rust epidemics can occur on susceptible cultivars in Australia. Despite rapid changes in virulence in the pathogen, damage has been minimized by timely release of new cultivars by the breeders. On three occasions there have been sudden changes in rust races in Australia that appear to have resulted from spores blown in from southern Africa (Luig 1985).

Leaf rust can be heavy in Australia when the weather is favorable but it causes less damage than stem rust. Again, resistant cultivars have helped to keep it under control.

Surprisingly, yellow rust did not appear in Australia until 1979 and in New Zealand until 1980 (Wellings and McIntosh 1982). So far it has been confined to the cooler areas of southeastern Australia. The single race that was originally identified was also present in Europe, suggesting that it was brought in by plane in some fashion. In 1980 stripe rust was more severe than either stem or leaf rust on susceptible cultivars (Luig 1985). Since the first appearance of yellow rust, new races that apparently originated by mutation have been identified.

1.4 Rust Control

The rusts are controlled primarily by either genetic resistance or the use of chemicals, and to a lesser extent by cultura! methods.

1.4.1 Cultural Methods

The objective of cultural methods for rust control is to break the life cycle of the rust, usually at a critical stage such as overwintering or oversummering. In some areas winter wheat is infected soon after emergence, by spores from nearby infected spring wheat, volunteer plants, or even late crops of winter wheat. Delaying planting, if it does not cause other problems, may prevent this early infection. Cultivation to remove volunteer plants from an earlier crop may also reduce sources of inoculum. Unfortunately, where winter wheat is seeded early for late fall or winter pasture, it may become infected and provide a source of early inoculum. On the other hand, in areas where rust inoculum normally arrives late in the growing season, planting early may allow a crop to mature before rust becomes serious. In some places where wheat is not grown over the summer, oversummering is the critical stage for the rust. Anything that can be done to eliminate susceptible hosts such as volunteer wheat or other susceptible species can help to control rust.

1.4.2 Chemical Control

Several fungicides are available that are effective, safe, and economic for use against rusts. In some cases, one spraying may be sufficient, but depending on the chemical, the weather, and the length of the growing season, two or more applications may be necessary. Chemicals are expensive and there is an added cost of application. Their use is economical only in areas such as western Europe where intensive cereal management is practiced, the same fungicide may control several diseases, and yields are high. In western Europe, in addition to the rusts, several other diseases such as powdery mildew (*Erysiphe graminis* f. sp. *tritici*), *Septoria* and *Fusarium* are present in varying amounts every year. Tram lines are often laid out for the wheels of tractors and sprayers to minimize crop damage from repeated applications of chemicals.

In some areas such as the Great Plains of North America, the rusts are a serious threat only infrequently, particularly in recent years. It is difficult and expensive to store chemicals and have them available quickly and in large quantities when they are needed. Predicting whether the use of a fungicide will be economic is difficult.

In the less developed countries, fungicides are usually imported and may not be readily available. Farmers may not be able to afford to buy chemicals or the equipment to apply them, and they may not be a feasible method of rust control.

Throughout the world, people are growing increasingly concerned about the use of chemicals and the potential danger of residues in food, of contamination of water supplies and of injury to the people applying them. Furthermore, in some cases pathogens have become resistant to fungicides. For these reasons, other methods of controlling rusts and other diseases are preferable.

1.4.3 Genetic Resistance

The development of resistant cultivars is the most effective method of biological control of the rusts. The use of resistant cultivars adds no extra cost to farmers. There are no chemicals to buy or additional cultural operations to carry out. Breeding for resistance is a very cost-effective procedure. Nevertheless, in many wheat-producing countries breeding for resistance appears to be a never-ending task that goes through a repeating cycle. Resistant cultivars are developed and released, their production increases rapidly, and new, virulent races of the rust appear. The breeding cycle then starts over again. As a result, wheat breeders are increasingly interested in types of resistance and methods of using resistance that will result in long-lasting control of the rusts.

1.5 Conclusion

Wheat is the world's most important crop. The rusts are present everywhere where wheat is grown and are among its most serious diseases. Consequently, the use of genetic resistance to control the rusts is of major importance throughout the world. The objective of this book is to examine the interaction of wheat and the wheat rusts in detail and to discuss the various ways in which genetic resistance can be used in the control of rust.

CHAPTER 2

The Wheat Rust Pathogens

2.1 Introduction

The wheat rust pathogens belong to the genus *Puccinia* of the family Pucciniaceae of the order Uredinales of the class Basidiomycetes (Littlefield 1981). In the Basidiomycetes, meiosis occurs in a basidium and results in the production of four, haploid, single-celled basidiospores. The order Uredinales includes the rust fungi, which are highly specialized plant pathogens. Depending on the taxonomist, the Uredinales are divided into two or more families based on the characters of the teliospores. By far the largest number of rust species belong to the Pucciniaceae and *Puccinia* is the largest genus in the family. Because the rust fungi are highly specialized and have narrow host ranges, identification of the hosts is an important aid in identifying a rust fungus.

2.2 Life Cycles

The life cycles of leaf and stem rust are similar and among the most complex of all fungi. Each has five spore types and both sexual and asexual stages (Fig. 2.1). The asexual stages occur on wheat and related grasses and the sexual stages on an alternate host, barberry (*Berberis* spp.) and *Mahonia* spp. for stem rust and *Thalictrum* spp. and some others for leaf rust. A rust that has all five spore stages is referred to as macrocyclic and one that has an alternate host is heteroecious.

As yet, only the asexual stage of yellow rust has been discovered. However, yellow rust produces basidiospores, the spore type that normally infects an alternate host. Despite considerable searching, an alternate host has not yet been identified. Making the connection between the rust stages on different hosts is not easy since the two hosts are normally unrelated taxonomically. The alternate host of yellow rust may yet be discovered. Alternatively, it may have become extinct or the pathogen may have lost its ability to infect it. Possibly the alternate host exists only in the center of origin of the rust, the Middle East or central Asia (Leppik 1970).

Some rusts, such as the flax rust (*Melampsora lini*), are autoecious, and complete all of their life cycle on one host. It is an interesting question as to how and why a life cycle with an alternate host evolved. Leppik (1967) believes that for stem rust *Berberis* (barberry) and *Mahonia* were the primary hosts from which the rust radiated out to the grasses. This greatly expanded the host populations available to it. The alternate hosts are of significance mainly in areas with cold winters where the asexual stage cannot survive but the telispores do. The resulting infection on barberry or *Mahonia* results in an earlier infection of the cereal host than would otherwise occur. In fact, the discovery of the alternate host of stem rust was aided by the observation that early infections often developed near infected barberry.

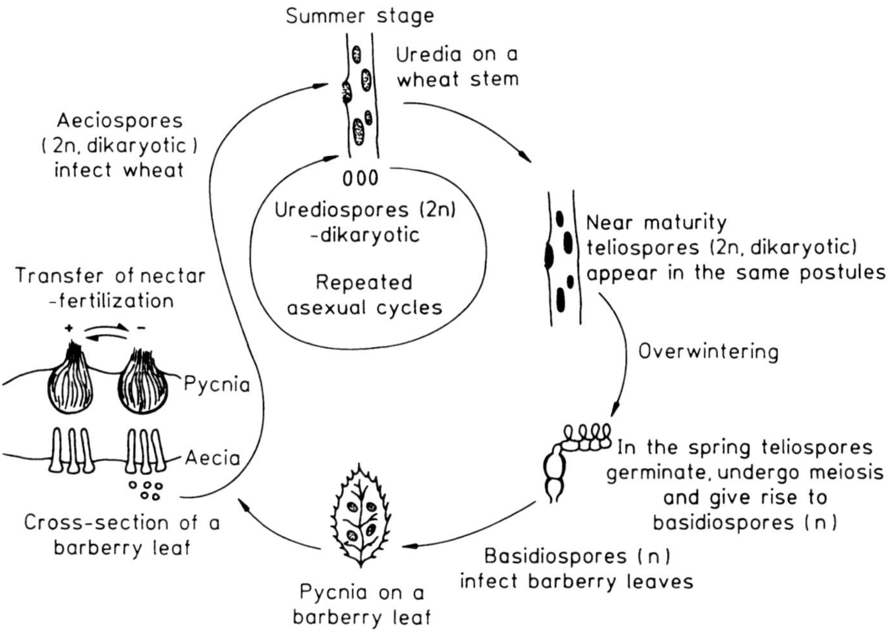

Fig. 2.1. The life cycle of wheat stem rust in North America

2.2.1 Wheat Stem Rust

The pustules of stem rust that are seen on wheat during most of its growing cycle are called uredia (singular uredium) and produce urediospores (Fig. 2.2). The terms uredinium and urediniospores are also used. The urediospores are dikaryotic (contain two genetically different nuclei), oblong in shape and reddish-brown in color. Uredia produce large numbers of spores for several weeks. When there is little wind, most spores remain within the crop canopy and cause reinfection. On windy days, many spores may become airborne and be carried long distances. For example, there is good evidence that spores have been carried from southern Africa to Australia on at least three occasions (Luig 1985). More commonly spores are blown from field to field over relatively short distances. Often stem rust survives a critical part of the year in only a limited area, e.g., an overwintering or oversummering area. As conditions become favorable and wheat crops begin to develop, the urediospores are carried out from such an area along what is often called a Puccinia path. In North America, stem rust overwinters in a limited area in the southern U.S.A. and Mexico. In the spring it begins to sporulate and the urediospores are gradually blown north along a *Puccinia* path of about 3000 km.

Urediospores germinate when free water is present on the wheat plants. Germ tubes grow until they reach a stoma. An appressorium is then produced over the stomatal opening, an infection peg pushes through it and a vesicle develops in the substomatal cavity (Staples and Macko, 1984) (Fig. 2.3). Infection hyphae grow from the vesicle and produce haustoria mother cells. Penetration pegs then push into the host cells and give rise to haustoria (Harder and Chong 1984). Depending largely

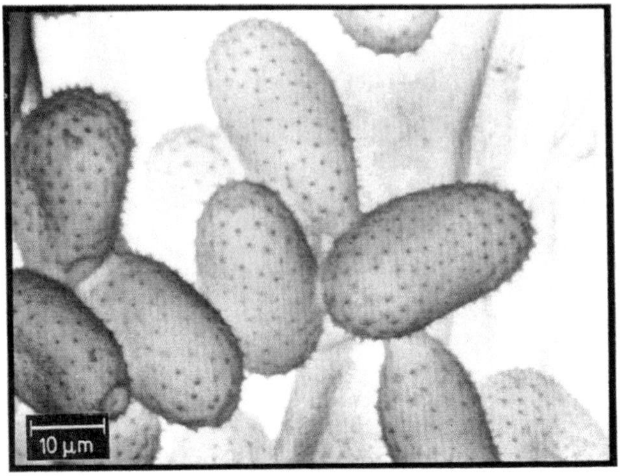

Fig. 2.2. Scanning electron micrograph of frozen/hydrated urediospores of wheat stem rust. (Photo courtesy of Dr. D.E. Harder, Agriculture Canada Research Station, Winnipeg)

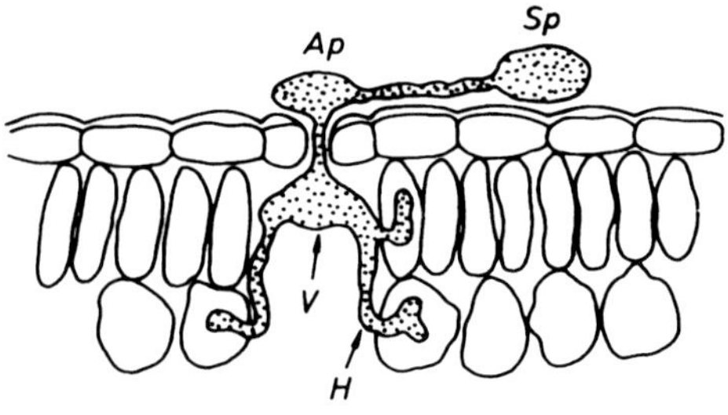

Fig. 2.3. Germination of a wheat stem rust spore (Sp) and the development of an appressorium (*Ap*), a substomatal vesicle (*V*), and a haustorium (*H*)

on temperature, infections usually become visible in 5–8 days and sporulation begins in 7–14 days. The shortest period occurs with the optimum temperature of about 30°C. Wheat stem rust can survive entirely in the asexual stage, and does in a number of countries where the alternate host does not occur.

Toward the end of the growing season the uredia begin to turn black as urediospore production ceases and black teliospores are produced in their place. The teliospores are borne on stalks, are two-celled and have blackish-brown caps (Fig. 2.4). These are the overwintering stage of the rust in cold climates. The teliospores are resistant to extremes of weather and germinate only after alternate periods of freezing and thawing, or wetting and drying. It has been very difficult to develop a reliable system to germinate teliospores under laboratory conditions. The teliospores are dikaryotic. As they mature the two nuclei become appressed and fuse, although fusion has not actually been observed because of difficulties in studying it (Harder 1984).

After overwintering, each cell of a teliospore may germinate to produce a basidium. The protoplast and nucleus migrate from the apical cell into the basidium. The nucleus then undergoes meiosis to produce four haploid cells (Mendgen 1984). These cells germinate to produce sterigma on the apices of which the basidiospores

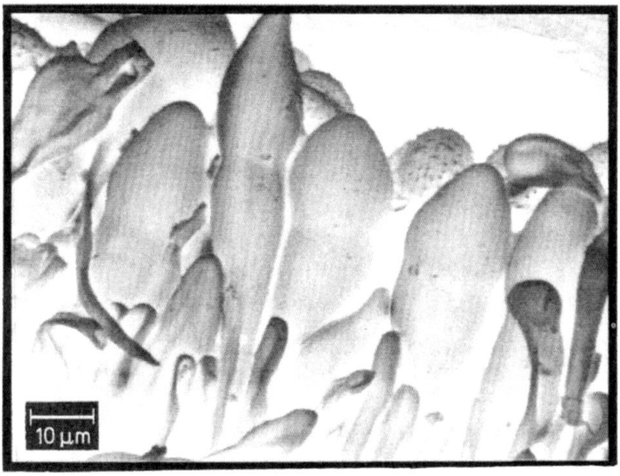

Fig. 2.4. Scanning electron micrograph of frozen/hydrated teliospores of wheat stem rust. (Photo courtesy of Dr. D.E. Harder. Agriculture Canada Research Station. Winnipeg)

are borne. The basidiospores are small. hyaline. and slightly oval-shaped. They are forcibly discharged into the air and carried to the alternate host. The basidiospores germinate rapidly. produce an infection peg and penetrate directly into epidermal cells (Roelfs 1985). Barberry genotypes may be either resistant or susceptible to particular stem rust genotypes.

The infection of barberry results in the development of subepidermal. flask-shaped pycnia on the upper leaf surface. About 7–14 days after infection the pycnia open and a viscous liquid (honeydew) containing pycniospores appears. The pycniospores are small and oval-shaped. and function as male gametes. The pycnia are of two mating types. + and −. and successful matings can occur only between opposite types. i.e.. the rust is heterothallic. The pycniospores are transferred from one pycnium to another by insects attracted to the honeydew. by splashing raindrops. or by leaves rubbing together in the wind. They can also be transferred artificially using a fine brush or glass rod. The pycnia contain long. thin. sterile hyphae. or paraphyses. which aid in the initial rupturing of the leaf epidermis and protrude through it. They also contain flexuous hyphae which function as female gametes and fuse with pycniospores of the opposite mating type. The function of the pycniospores was not determined until the work of Craigie (1927. 1931). After fusion. the nucleus of a pycniospore migrates through the flexuous hyphae to the aecial primordium. a process that takes about 20–25 h. The mycelium then develops into an aecium on the underside of the barberry leaf. The mycelial cells are dikaryotic. A single aecium can contain many aecial horns. cylindrical structures that flare open at the tip. Since a number of fertilizations can occur in a pycnium. the resulting aecium will be genetically heterogeneous. Aeciospores from a single aecial horn are usually of one genotype. indicating that they are the product of a single fertilization (Roelfs 1985). The aeciospores are produced in long chains and are forcibly discharged when aecia are wetted and dried. The aeciospores are roundish and covered with a layer of fine knobs. They are produced in large numbers and are carried by the wind to nearby wheat fields where they can produce heavy, early infections. The spores can probably be carried over long distances but heavy infections occur in the immediate vicinity of the barberry. The infections on wheat develop into uredia, thus completing the life cycle.

In many areas barberry is not present (e.g., much of Australia) or the sexual cycle on barberry is of little importance. In the central *Puccinia* path in North America, barberry has been largely eradicated. The rust is then forced to survive in its asexual stage in the overwintering area in the southern states, which it does all too well. Since there is no dormant stage, the rust survives stress periods, cold or heat, as living mycelium in living tissue of wheat plants or some other hosts, such as perennial grasses.

2.2.2 Wheat Leaf Rust

The life cycle of wheat leaf rust is very similar to that of stem rust. The alternate hosts are primarily *Thalictrum* spp. but several other genera function in limited areas (Samborski 1985). In most areas the alternate hosts do not appear to play a major role in the life cycle by initiating early infections in the spring. However, at least in some areas the sexual cycle is probably important in the production of new combinations of virulences by genetic recombination (Samborski 1985).

The uredia of leaf rust are small, round and orange-red in color, and sporulate primarily on the upper leaf surface. The urediospores are small and almost perfectly spherical (Fig. 2.5). They are easily distinguishable from those of stem rust by their smaller size and spherical shape compared to the oblong shape of stem rust. The teliospores of leaf rust are also distinguishable from those of stem rust, being flatter and broader at the tip, rather than rounded.

2.2.3 Wheat Yellow Rust

Since no alternate host has been identified for yellow rust, only three spore stages are known, urediospores (Fig. 2.6), teliospores, and basidiospores. Attempts to infect possible alternate hosts using basidiospores have not been successful. The urediospores and teliospores of yellow rust are very similar to those of leaf rust and, therefore, easily distinguishable from those of stem rust.

The life cycle of yellow rust involves repeated cycles of the asexual uredial stage. A major difference between yellow rust and the other two rusts is that a single infection on a leaf can produce a long stripe containing many uredia. The urediospores are an orange-yellow color. On resistant plants, there are frequently large areas of chlorotic or necrotic tissue. Yellow rust can survive periods of stress, cold in the winter and heat in the summer, as mycelium in tissue of living wheat plants. However, if all above ground parts of the plant are killed, the rust will not survive. Thus, in particularly severe winters yellow rust does not survive in northwestern Europe, although in many years it does survive.

Fig. 2.5. Scanning electron micrograph of frozen/hydrated urediospores of wheat leaf rust. (Photo courtesy of Dr. D.E. Harder, Agriculture Canada Research Station, Winnipeg)

Fig. 2.6. Scanning electron micrograph of frozen/hydrated urediospores of wheat yellow rust. (Photo courtesy of Dr. D.E. Harder, Agriculture Canada Research Station, Winnipeg)

2.2.4 Chromosome Numbers

The chromosomes of the wheat rusts are very small and it has been difficult to make clear cytological preparations for chromosome counts. Most counts have been done on mitotic divisions in basidiospores or the apical cells of hyphae in wheat leaves.

McGinnis (1953) reported that there were six chromosomes in haploid nuclei of *Puccinia graminis* on *Agropyron trachycaulum* and Valkoun and Bartoš (1974) reported the same number for *P. recondita*. Goddard (1976a) found six chromosomes in haploid nuclei of *P. striiformis*. However, Wright and Lennard (1978) found only three chromosomes in most cells and suggested that Goddard (1976a) might have counted areas of condensed chromatin. Several authors have reported that at prophase the chromosomes in several *Puccinia* species are connected in rings (e.g., McGinnis 1956). The basic chromosome number in *Puccinia* is probably three, with a number of species being tetraploids with n = 6. Several species have four chromosomes, a number that may have arisen by duplication of one chromosome (McGinnis 1956).

McGinnis (1956) studied a number of species of *Puccinia* and suggested that whether they were homothallic or heterothallic depended on chromosome number. Homothallic species had n = 4, while heterothallic species had either n = 3 or n = 6.

2.3 Axenic Culture of the Rusts

In nature the rusts are obligate biotrophs. They obtain their nutrients from living wheat cells and die if their host dies. Despite many attempts, none of the wheat rusts was cultured on an artificial medium until P.G. Williams et al. (1966) were successful with stem rust (see reviews by Maclean 1982; Williams 1984). Since then, leaf rust and yellow rust have been cultured as well.

Although methods are now available for culturing wheat rusts, there is no standard procedure which will guarantee the successful culture of all genotypes or can be used routinely to produce vigorous cultures. A single rust genotype cultured under one set of conditions often gives quite variable results. Genotypes differ greatly in their ability to grow in culture. Thus, axenic culture has not become a standard procedure for the rusts.

2.4 Classifying Infection Types (ITs)

Starting about 1910, E.C. Stakman and his co-workers at the University of Minnesota studied the interaction between stem rust cultures and wheat cultivars. They developed a system of designating infection types for uredia on seedling leaves on a 0 to 4 scale with an extra class, designated X, for heterogeneous or mesothetic infections (Stakman et al. 1962) (Table 2.1). The system has been widely used for stem and leaf rust. Two additional classes are sometimes added, particularly for leaf rust. They are heterogeneous or pattern types. A Y infection type indicates that the uredia are variable in size and are largest and often most frequent at the tip of the leaf blade. In a Z infection type the pattern is reversed, the uredia are larger and more frequent at the base of the leaf blade. In classifying leaf rust, allowance must be made for the much smaller size of the uredia. The methods of producing rust infections under controlled conditions are described in Chap. 3.

For both rusts, the infection types make up a continuum and there is no clear separation between types. For example, uredia may be somewhat larger than typical for infection type 1 but have its typical yellow chlorosis. Are they then IT1 or IT2? Each worker will make his or her own decision, based on experience. For conven-

Table 2.1. Description of infection types used in classifying the reactions to stem rust on seedling wheat leaves. (After Stakman et al. 1962)

Class	Infection type[a]	Description of symptoms
Immune	0	No signs of infection to the naked eye but minute flocks may be visible under low magnification
Very resistant	0;	No uredia, but distinct flecks of varying sizes, usually a chlorotic yellow but occasionally necrotic
Resistant	1	Small uredia surrounded by yellow chlorotic or necrotic areas
Moderately resistant	2	Small to medium-sized uredia, typically in a dark green island surrounded by a chlorotic area
Mesothetic or heterogeneous	X	A range of infection types from resistant to susceptible scattered randomly on a single leaf; caused by a single isolate not a mixture
Moderately susceptible	3	Medium-sized uredia, usually surrounded by a light green chlorosis
Susceptible	4	Large uredia with a limited amount of chlorosis; may be diamond-shaped

[a] Pluses and minuses are used to indicate variations from the size of uredia typical of a particular infection type as follows: = uredinia much smaller than typical and at the lower limit for the infection type, – uredia smaller than normal, + uredia larger than normal, + + uredia much larger than typical and at the upper limit for the infection type

ience, infection types 0, 0; 1, 2 and X are normally considered resistant and 3 and 4 are considered susceptible. However, in a segregating population from a cross such a distinction will have no validity if there is a continuous range of infection types. Roelfs (1984) classifies 0 to 3 as low infection types and 4 as high.

Other systems of classifying infection types have been developed. At the University of Sydney, the infection types for stem rust based on Stakman et al. (1962) are classified into four groups, necrotic infection types, green island infection types, chlorotic infections, and heterogeneous infection types (Luig 1983). Within each group there are 9 to 12 ITs which are numbered consecutively from 1 to 41.

Browder and Young (1975) proposed a system for coding infection types for the cereal rusts that used two digits and a letter. The first digit indicated the relative amount of sporulation and the second the relative lesion size, both on a 0–9 scale. The letter indicated the nature of tissue damage as follows:

C – chlorotic tissue associated with typical lesions.
D – the entire infected leaf dead, apparently caused by rust infection.
G – a green island associated with typical type 2 lesions.
K – Kenya-type chlorotic-necrotic tissue associated with typical lesions.
N – necrotic tissue associated with typical lesions.
P – pale green tissue surrounding the sporulation of typical lesions.
W – white tissue associated with typical lesions.
X – classic X type.
Y – classic Y type.
Z – classic Z type.
– – no data available.

To be efficient, the system requires accurate control of inoculum densities and environmental conditions so that uniform infection occurs. Otherwise the size and sporulation of uredia are affected by their density on the leaf. The system has been used primarily with leaf rust.

Yellow rust has been treated somewhat differently than other rusts because a single infection can result in a long stripe on a leaf. Chlorosis and necrosis can be very extensive without any sporulation occurring or there may be many uredia in a single stripe. Gassner and Straib (1932) developed a detailed set of infection types. Various other authors have used simplified classifications which differ mainly in the way the nonsporulating infection types are classified. Categories 1, 2, 3 and 4 are essentially the same in most systems. Lewellen et al. (1967) used the following infection types:

$0^{v}0$	= nearly immune to immune but minute chlorotic flecks usually present.
00	= small, symmetrical, necrotic flecks.
0–	= larger necrotic flecks, usually nonsymmetrical.
0	= larger areas of necrosis but no pustules.
1–	= similar to the 0 type but eventually forming very small pustules.
1	= necrosis with small pustules.
2	= necrosis and chlorosis with larger pustules.
3	= no necrosis but chlorosis with normal pustules.
4	= no necrosis or chlorosis and normal pustules.

However, Line et al. (1970) used a 0 to 9 scale with the first six classes considered as avirulent and the last four as virulent.

2.5 Variability in the Rust Pathogens

2.5.1 Formae Speciales

Within a rust species, the first level of variability is the formae speciales. They are defined by their ability to attack a particular species or a group of related species but do not differ in morphology. Usually they are named after the most common or most important species that they attack. J. Ericksson did extensive studies on the formae speciales and Ericksson and Henning in 1896 proposed four formae speciales in *P. graminis*, two in *P. dispersa* (*P. recondita*) and three in *P. glumarum* (*P. striiformis*) (see Schafer et al. 1984).

The formae speciales within a species have differentiated from one another to various degrees. For example, *P. graminis* f.sp. *tritici* crosses readily with f.sp. *secalis* but is very difficult to cross with f.sp *avenae* (Johnson 1949). In general, hybrids tend to be able to attack fewer genotypes within a host species but have a wider host range than their parents.

2.5.2 Physiologic Races

Between 1910 and 1920, Stakman and his coworkers at the University of Minnesota tested a number of wheat lines with a number of cultures of stem rust. They discovered that within wheat stem rust there were a number of different biologic forms (physiologic races) that were stable but differed in virulence on particular host lines (Stakman et al. 1919; Stakman and Levine 1922). Twelve lines were selected as a standard set of differential hosts on which to identify rust cultures. All cultures that gave the same pattern of virulence and avirulence on the standard differentials were considered to be a single physiologic race. They could, however, differ in virulence on other host lines or in other characters. The set of 12 lines became known as the standard or international or Stakman differentials and is still in use. It includes cultivars from three ploidy levels (Table 2.2). The inclusion of cultivars from species with three different ploidy levels made it more difficult to determine the genes for resistance carried by them. Even now they have not been completely analyzed, and, of course, they carry resistant alleles at only some of the known loci for rust reaction. The chromosome numbers of the wheat species were just being determined at the time the set of differentials was being selected. Eventually it was discovered that several differentials gave reactions similar to others. The genes for rust resistance are named after the first letters of the common names for the diseases, *Sr* for stem rust, *Lr* for leaf rust, and *Yr* for yellow rust. Numbers are used to specify specific genes and small letters for alleles, except that letters are used for the temporary designation of genes whose relationship to the other known genes is not completely worked out.

Some workers use an abbreviated set of the standard differentials (e.g., Marquis, Reliance, Acme, Einkorn, and Vernal in Australia), plus supplementary differentials (Sect. 2.5.3). In North America, for the asexual rust population it is possible to predict the standard race number from the response of lines carrying *Sr5, Sr7b, Sr9d,* and *Sr9e* (Roelfs 1985).

Similar specialization was discovered in leaf rust by Mains and Jackson (1926), who developed a set of 11 differentials, all hexaploids. Three of these were eventually dropped (Johnston and Mains 1932) and the remaining eight became accepted internationally. They are Brevit (*Lr2c*), Carina (*Lr2b*), Democrat (*Lr3*), Hussar (*Lr11*), Loros (*Lr2c*), Malakoff (*Lr1*), Mediterranean (*Lr3*), and Webster (*Lr2a*).

Table 2.2. The standard or international differential set of cultivars for identifying wheat stem rust cultures. (Based on Luig 1983; Roelfs 1985)

Species and cultivars	Known genes for resistance
Triticum monococcum (2n = 14, AA)	
Einkorn	*Sr21*
Triticum timopheevii (2n = 28, AAGG)	
Khapli	*Sr7a, Sr13, Sr14*
Triticum turgidum (2n = 28, AABB)	
Acme	*Sr9g* plus gene A[a]
Arnautka	*Sr9d* plus gene B[a]
Kubanka	*Sr9g* plus gene C[a]
Mindum	*Sr9d* plus gene B
Spelmar	*Sr9d* plus gene B
Vernal	*Sr9e*
Triticum aestivum (2n = 42, AABBDD)	
Kota	*Sr7b, Sr18, Sr28, SrKt2*
Little Club	*SrLC*
Marquis	*Sr7b, Sr18, Sr19, Sr20, SrX*
Reliance	*Sr5, Sr16, Sr18, Sr20*

[a] Genes A, B and C condition mesothetic or X infection types but have not been identified further.

The reactions of three of the cultivars, Carina, Brevit, and Hussar, proved to be particularly sensitive to temperature and it was suggested that they be dropped (Chester 1946). Basile (1957) developed new rust keys based on the remaining five differentials. While this resulted in more consistent reactions, it also reduced the amount of information available. A committee of North American Wheat Leaf Rust Research Workers recommended retaining the standard differentials and adding some supplementary differentials (Loegering et al. 1959). Most workers continue to use the full set and report their results according to both keys (Samborski 1985).

Hungerford and Owens (1923) were the first to report the occurrence of different "strains or specialized forms" in yellow rust. Researchers in Braunschweig, West Germany, did extensive studies on physiological specialization and developed a differential set that included 12 wheat cultivars, 2 barleys and 1 rye (Stubbs 1985). Consistent race identification has been more difficult with yellow rust than with the other two rusts. Fuchs (1960) found it necessary to group races that did not differ greatly. Fuchs and Zadoks initiated an international race survey (Zadoks 1961). Johnson et al. (1972) reviewed the identification and naming of physiologic races of yellow rust. As a result of their review, a basic set of seven differentials for worldwide use was established. The seven are Chinese 166 (*Yr1*), Heines Kolben (*Yr6*), Lee (*Yr7*), Moro (*Yr10*), Strubes Dickkopf, Suwon 92/Omar, and Vilmorin 23 (probably *Yr3*). Extensive use of supplementary differentials is made in various parts of the world (see Sect. 2.5.3).

For all three rusts, physiological races are identified solely on the basis of their virulence on a set of genotypes. Races do not differ consistently in other characters such as spore size, shape, or color.

2.5.3 Supplementary Differentials

With all three rusts, as soon as a standard set of differential cultivars for identifying races was established it was found that further subdivisions could be made by using additional differentials carrying either genes for resistance not present in the standard sets or new combinations of genes. Often these were resistant cultivars produced by wheat breeders. Therefore, the presence or absence of virulence on them was of particular importance.

The rust pathologists in Australia have developed a system of identifying and naming races of stem rust using important commercial cultivars as supplementary differentials (Watson and Luig 1963). For example, when Eureka, which carries gene *Sr6* for stem rust resistance, was released in 1938, virulence quickly appeared in the rust population. A-1 was added to a standard race designation to indicate virulence on Eureka. Thus, 21-1 was standard race 21 but with virulence on Eureka (*Sr6*). Initially it was not always known which gene for resistance had been overcome. Sometimes ANZ was added to indicate that the strain was identified in Australia and New Zealand, e.g., 21 ANZ-1, but this has often been omitted in recent years. As virulence appeared on new cultivars, additional numbers was added (Table 2.3). Virulence on several cultivars or lines arose despite the fact that they had not been grown commercially in the field and, thus, there should have been no selection pressure in favor of them (Luig 1983).

Table 2.3. Supplementary differentials used to identify stem rust isolates in Australia. (Luig) 1983; McIntosh, personal communication)

Supplementary differential	Wheat cultivar or line[a]	*Sr* gene
1	Eureka	*Sr6*
2	Gabo	*Sr11*
3	Gamenya	*Sr9b*
4	Mengavi	*Sr36*
5	Spica	*Sr17*
6	Oxley	*Sr8a*
7	Thew[b]	*Sr15*
8	Festiguay	*Sr30*
9	*Agropyron intermedium* derivative[b]	*SrAgi*
10	Entrelargo de Montijo[b]	*Sr51209*
11	Barleta Benvenuto[b]	*SrBB*
12	Coorong triticale	*Sr27*
13	Satu triticale	*SrSatu*

[a] Often several related cultivars can be used but only the key one is given.
[b] Virulence for these cultivars or lines arose despite the absence of selection pressure resulting from their use in commercial production.

Supplementary differentials have been developed in other major wheat-growing areas and for the other wheat rusts. The major purpose has been to provide specific information needed by wheat breeders. Rust populations in different areas can differ substantially, often depending on the genes for resistance carried by the cultivars in commercial production. Isolates from different areas may key out to be the same race when tested on a standard set of differentials but may differ markedly when supplementary differentials are used. It cannot be assumed that standard races from different areas are the same.

For leaf rust, the North American Wheat Leaf Rust Research Committee recommended a system that retained the standard differentials but used supplementary differentials specific for North America (Loegering et al. 1959). Sixteen cultivars were under test as potential supplementary differentials. The Committee suggested including some lines resistant to all current races in order to detect new virulences. It was agreed that races would be designated by a standard race number, NA for North American supplementary differentials, and a number indicating the virulence pattern on the supplementary differentials, e.g., 9-NA-1. Later the Committee proposed the use of a set of four supplementary differentials, designated as NA61 and made up of Lee (*Lr23*), Sinvalocho, Waban, and Westar. Finally, Westar was dropped and Dular and Exchange were added (Young and Browder 1965).

The Australian workers used the same system for leaf rust that they used for stem rust, i.e., standard race designations with numbers added to indicate virulence on specific supplementary differentials (Watson and Luig 1961).

For yellow rust, Johnson et al. (1972) recommended a set of eight supplementary differentials for use in Europe: Hybrid 46 (*Yr4b*), Reichersberg 42 (*Yr7*+?), Heines Peko (*Yr6*+?), Nord Desprez, Compair (*Yr8*), Carstens V, Spaldings Prolific, and

Heines VII (*Yr2*). Three of these are also used in India, along with a larger set of local cultivars and lines (Stubbs 1985). Another group of cultivars is used in the north-western United States and a very large group of cultivars in China (Stubbs 1985).

2.5.4 Race Identification Using Other Sets of Differentials

A major problem with the three sets of standard differentials for the three rusts was that initially little was known about the genes for resistance that they carried. When genetic data were obtained, it often turned out that some cultivars were similar genetically and were of little value in distinguishing races. The stem rust set had the added problem that species at three different ploidy levels were involved. As genetic information was obtained and the gene-for-gene theory developed (see Chap. 6), there was increasing interest in using genetically known stocks, and particularly near-isogenic lines, in race identification. Near-isogenic lines are usually produced by backcrossing and have a common genetic background except for resistance genes and genes linked to them.

2.5.4.1 Near-Isogenic Lines and Virulence Formulas

In Canada, a set of 23 single-gene differentials is used to identify stem rust cultures (Green 1981). Thirty-seven loci for stem rust resistance have been named and three have two or more alleles, but not all are used. Some of the differentials were derived from backcrosses to susceptible cultivars such as Marquis and Prelude, but others are cultivars or lines known to carry single genes (Table 2.4). Thus, the resistance genes are not in a common genetic background and some lines may carry other genes for resistance. Three of the standard differentials, Marquis, Mindum, and Einkorn, are also included. They, along with some single-gene lines, are used to identify the standard race number for rust cultures. Four commercial cultivars, Manitou, Selkirk, Sinton and Neepawa, are also tested to determine which races may be potential threats in Canada. Each culture tested is given a standard race number and a virulence formula (effective genes/ineffective genes). As races have been identified over time using the new system, they have been given C numbers. Thus, C1, identified in 1964, is: C1(17) 5,6,7a,9a,9b,9e,10,11,13,17/7b,8a,9d,14,15,16, i.e., it is standard race 17. Not all of the single-gene lines were used in the identification of C1. The formula method provides very precise information about the pathogenicity of rust races. It has the advantage that it is open-ended, i.e., new single-gene lines can be added. However, the virulence formulas can become very long.

The development of a set of lines of Thatcher carrying single genes for leaf rust resistance led to their use throughout the world as differential lines (Samborski 1985). Nineteen lines were used initially and new lines can be added as they are produced (Samborski 1980). Races are designated by virulence formulas but are not given a race designation as is done for stem rust.

Table 2.4. Cultivars and lines of wheat carrying single genes for resistance used to identify stem rust cultures in Canada. (Green 1981)

Genes for resistance	Typical resistant infection type	Cultivar or line
Sr5	0	Prelude*6/Reliance
Sr6	0;	Mida-McMurachy-Exchange/ 6*Prelude
Sr7a	0; to 3CN	Na101/6*Marquis
Sr7b	2	Chinese Spring/Hope (C.I. 14165)
Sr8a	2	Chinese Spring/Red Egyptian (C.I.14167)
Sr9a	2	Chinese Spring/Red Egyptian (C.I. 14169)
Sr9b	2	Prelude*4//Marquis*6/Kenya 117A
Sr9d	2	H-44-24/6*Marquis
Sr9e	12	Vernstein W3196
Sr10	X-	Marquis*4/Egypt Na95//2*W2691
Sr11	1+	Chinese Spring/Timstein C.I. 14171
Sr13	2	Prelude*4/Marquis*6/Khapstein
Sr14	2	W2691*2/Khapstein
Sr15	X	Prelude*2/Norka, R.L. 1888
Sr17	0;1	Prelude/8*Marquis*2//Esp 518/9
Sr22	2	Marquis*6//3*Stewart/R.L. 5244
Sr24	2	Agent (C.I. 13523)
Sr26	2	Eagle
Sr27	0;	WRT 238-5
Sr29	2	Prelude/8*Marquis//Etoile de Choisy
Sr30	2	Webster
Sr36	0;1	Prelude*4/NHL II.64.62.1
Sr37	0;	W3563

2.5.4.2 Coded Sets of Differentials

In the United States, a different system has been developed for stem rust that also uses single gene lines (Roelfs et al. 1982b; Roelfs 1985). The lines carry *Sr5, Sr6, Sr7b, Sr8a, Sr9a, Sr9d, Sr9e,* and *Sr15* in a Chinese Spring background and *Sr9b, Sr10, Sr13, Sr15,* and *Sr36* in the background of W2691, a highly susceptible line developed at the University of Sydney. Four other testers are Vernstein (*Sr9e*), Combination VII (*Sr13* and *Sr17*) and Triumph 64 (*SrTmp*). Thus, the system uses most of the genes that are also used in Canada. To designate a race, a coding system is employed. Twelve of the testers are divided into three sets of four (Table 2.5). The first set includes the genes *Sr5, Sr7b, Sr9d,* and *Sr9e* which were the most important genes in the standard differentials for identifying North American races. The other two sets include some of the more useful genes for race identification that have been identified in recent years. When four lines are classified for resistance or susceptibility there are 16 possible combinations. These are coded from B to T, omitting the vowels (Table 2.5). The pathogenicity of a race is coded using three letters, each indicating its pathogenicity on one set of four lines. Code BBB would indicate that

Table 2.5. The coding system used in the United States to designate stem rust races and indicate their pathogenicity. (Roelfs et al. 1982b)

	Wheat lines carrying			
Set 1	*Sr5*	*Sr9d*	*Sr9e*	*Sr7b*
Set 2	*Sr11*	*Sr6*	*Sr8a*	*Sr9a*
Set 3	*Sr36*	*Sr9b*	*Sr13*	*Sr10*
Code[a]		Host reaction		
B	R[b]	R	R	R
C	R	R	R	S
D	R	R	S	R
F	R	R	S	S
G	R	S	R	R
H	R	S	R	S
J	R	S	S	R
K	R	S	S	S
L	S	R	R	R
M	S	R	R	S
N	S	R	S	R
P	S	R	S	S
Q	S	S	R	R
R	S	S	R	S
S	S	S	S	R
T	S	S	S	S

[a] Three letters designate a race. The first letter indicates the reaction of the lines in set 1, the second letter set 2 and the third letter set 3.
[b] R = resistant, S = susceptible.

all 12 tester lines are resistant and TTT would indicate that all 12 are susceptible. From the data, it is also possible to determine the standard race number and it is used in the race designation. In recent years the most common stem rust race has been 15-TNM, standard race 15, to which the lines in set 1 are all S (code T), the lines in set 2 are S, R, S and R (code N), and the lines in set 3 are S, R, R and S (code M). The system has the advantage that the race designations are simple, but it is necessary to have the table of codes available to interpret the letter (unless a person has a photographic memory). New sets can be added at any time without greatly lengthening the designation.

2.5.4.3 Mathematical Systems for Race Designation

Habgood (1970) proposed a decanary system for designating races of plant pathogens. Johnson et al. (1972) recommended its use for yellow rust. In the system, a considerable amount of information is coded in a single number. The differential hosts are listed in a fixed order and assigned decanary numbers which are the successive powers of 2 (Table 2.6). The reactions of the hosts to an isolate are then assigned binary scores, 0 for resistant, and 1 for susceptible. Only the susceptible hosts are used to determine a race number for the isolate. The race number is obtained by adding the decanary values for the susceptible hosts. In the example, the sum of the values for the four susceptible hosts is 86. Given the race designation, it

Table 2.6. The system of designating rust races as proposed for yellow rust by Johnson et al. (1972), illustrated using results for a single rust isolate

Host line	A	B	C	D	E	F	G	H
Decanary value	$2^0=1$	$2^1=2$	$2^2=4$	$2^3=8$	$2^4=16$	$2^5=32$	$2^6=64$	$2^7=128$
Reaction to isolate	R	S	S	R	S	R	S	R
Binary score	0	1	1	0	1	0	1	0
Decanary values for susceptible lines	-	2	4	-	16	-	64	-
Race designation			86 (2+4+16+64)					

is possible to determine which host lines are susceptible to the race, as follows. From 86 subtract the decanary value that is just below it, i.e., 64 (86–64=22). In the same way, from 22 subtract 16=6, from 6 subtract 4=2, and from 2 subtract 2=0. Race 86 is virulent on the hosts with decanary values of 1, 4, 16, and 64, i.e., B, C, E, and G. Every combination of resistant and susceptible reactions on the differential hosts results in a unique number. Johnson et al. (1972) recommended that European races of yellow rust be given a number based on the standard differentials, an E for European and another number based on the European supplementary differentials, e.g., 36 E 128. The disadvantage of the decanary system is its complexity. It is not simple to uncode the numbers and determine which hosts are susceptible. The advantage is that it is open-ended – new hosts can be added at any time although the addition of a new, susceptible host will change the previous race designations.

Fleischmann and Baker (1971) proposed a binomial system of notation and Gilmour (1973) an octal notation.

2.5.5 Races, Biotypes, Cultures, Isolates

The original name "physiologic races" is in a sense a misnomer. As defined, races are types differing in pathogenicity, i.e., pathogenic races. A race is a specific combination of virulence and avirulence on a defined set of differentials. It is not a single genotype. As the work with supplementary differentials showed, a race can comprise numerous genotypes differing in virulence on additional differentials. A race can be heterogeneous for other characters such as spore color, but few additional genetic markers are known, although in recent years isozymes have been used (Burdon et al. 1982; Newton et al. 1985) and restriction fragment length polymorphisms (Worland et al. 1987).

When additional host genotypes were employed to subdivide a race, the term biotype was often used to designate the subunit. However, a subunit could often be further subdivided by the use of additional host genotypes. For example, stem rust race 15 is avirulent on Lee. A type was found that was virulent on Lee and was designated biotype 15B. Then 15B was found to include at least five types which were designated biotypes 15B-1 to 15B-5. Biotype 15B-1 is avirulent on Langdon durum but a type virulent on Langdon was discovered and designated 15B-1L. It has been important in North America for many years. The key point is that there is no end to the number of divisions that can be made.

As Roelfs (1984) pointed out, a biotype is defined as "a population of individuals of the same genotype". Luig (1985) uses race to refer to "biotypes of the rust pathogens that have been identified on the international differentials" and strain to refer to "biotypes that have been separated on supplementary differentials."

In recent years, some authors have questioned the continued use of the term race. Day (1974) suggested that in the future race designations would become largely unnecessary. Browder (1971) concluded that, "The classic pathogenic race concept is inadequate" because the information it give applies only to specific sets of differential cultivars, components within a race may differ in virulence on other cultivars, and a race key is necessary to provide information about races. Vanderplank (1983) recommended throwing out races completely because there is a potentially unlimited number of races if new differentials are added and because races as defined are fixed entities, whereas what is really of interest is gene changes in rust populations.

As pointed out by Caten (1987), the race concept is of value where recombination of genes within a population is limited, and the population is made up of a relatively small number of fixed genetic units. This is true for many rust populations that reproduce asexually. It is, then, the relative frequency of these units that is important to the wheat breeder. The continued use of races makes it possible to follow the evolution of a rust in a given area. However, if a standard set of differential cultivars is used, the emphasis is on cultivars that may be of little interest to the plant breeder. This has led to the use of supplementary differentials and single gene differentials, and to other systems of race nomenclature as already described. What is often of interest to plant breeders is the frequency of virulence on specific genes for resistance, changes in those frequencies and in the frequencies of particular combinations of virulence genes. Wolfe and Schwarzbach (1975) recommended "virulence analysis" for the cereal mildews. Such a system is open-ended since new differentials can be added as new genes for resistance become important. As is usual in population genetics, the analysis is based on gene rather than genotype frequencies.

The methods of studying rust populations are still evolving. There is still value in identifying standard races so as to maintain continuity with previous work and to measure evolution of rust populations. As more sets of near-isogenic lines become available, increasing use will be made of them.

In studying rust collections, the terms culture and isolate have tended to be used interchangeably. If there is a distinction, then an isolate is more likely to be derived from a single uredium or a single urediospore and, therefore, to be a pure line. A culture often originates from a field collection and its purity is less certain.

2.6 The Origin of Pathogenic Variability

2.6.1 Sexual Recombination

After Craigie (1927) had described the sexual cycle of stem rust, Newton et al. (1930a) both selfed and crossed several stem rust races. One race appeared to be homozygous but others gave rise to one or more types different than the parent.

Crosses between races gave rise to types different from both parents and in some cases not previously known. From this paper and others that followed, it soon became apparent that avirulence was often dominant. Thus, the presence of genes for virulence could be hidden in heterozygous loci. Selfing of a rust isolate could result in types more virulent than the parent. For example, a race that was avirulent on the genes for stem rust resistance, *Sr5* and *Sr6*, but heterozygous at both loci for pathogenicity, on selfing would give rise to eight new genotypes, and four new phenotypes, i.e., *P5p5 P6p6* selfed (p = virulent, P = avirulent) gives:

P5P5 P6P6 – avirulent on both *Sr5* and *Sr6*
P5P5 P6p6 – avirulent on both
P5P5 p6p6 – avirulent on *Sr5*, virulent on *Sr6*
P5p5 P6P6 – avirulent on both
P5p5 P6p6 – avirulent on both (parental genotype)
P5p5 p6p6 – avirulent on *Sr5*, virulent on *Sr6*
p5p5 P6P6 – virulent on *Sr5*, avirulent on *Sr6*
p5p5 P6p6 – virulent on *Sr5*, avirulent on *Sr6*
p5p5 p6p6 – virulent on both.

Five of the genotypes are virulent on more resistance genes than the parent.

For stem and leaf rust, the sexual cycle can be an important source of new combinations of genes for virulence wherever the alternate host occurs. Roelfs and Groth (1980) compared an asexually reproduced population of wheat stem rust from an area east of the Rocky Mountains in the United States with a sexually reproduced one from Idaho and Washington. Isolates were tested on 16 lines carrying single genes for stem rust resistance. In the sexual population, 100 distinct phenotypes were obtained from 426 isolates while in the asexual population 17 phenotypes were identified in 2377 isolates. However, the sexual population averaged only six genes for virulence per isolate compared to about ten for the asexual population. Sexual recombination results in the production of many more combinations of virulences and also, undoubtedly, considerable variation in fitness. In the asexual population, competition has resulted in the survival of a small number of the fittest genotypes. In areas where the alternate host is present and the sexual cycle occurs, sexual recombination can result in the production of new genotypes for virulence. In North America, the very important race 15B of stem rust was first found near barberry.

2.6.2 Mutation

Genes for virulence in the rust pathogens are generally recessive. Recessive mutations occur much more frequently than dominant ones, although even their frequency is low, perhaps of the order of 1×10^{-5} or 10^{-6}. Since the rusts are dikaryotic, a double mutant will be needed and its frequency will be of the order of 1×10^{-10} to 1×10^{-12}. However, urediospores are produced in immense numbers. Rowell and Roelfs (1971) estimated that an acre (0.4 ha) of wheat with a stem rust severity of 10% can produce a trillion spores (1×10^{12}). Similarly, Parlevliet and Zadoks (1977) estimated that 1 ha of wheat with 1% of the leaf area occupied by leaf rust uredia could produce 1×10^{11} spores. Thus, if a field of a new cultivar carrying a single gene for resistance is surrounded by fields of susceptible cultivars, large numbers of spores will be blown into it and a few will be virulent mutants. If only the mutants can infect the plants and sporulate, the extreme selection pressure will result in their rapid increase. Frequently in wheat-producing areas, as a new, rust-resistant cultivar becomes widespread, selection pressure results in the rapid spread of a new, virulent race. The evolution of rust populations in the field is often driven by the resistant cultivars produced by the wheat breeders and grown by farmers. Johnson (1961) referred to this as "man-guided evolution in plant rusts."

Luig (1983) did an extensive review of the literature on race changes in stem rust in the major wheat-growing areas of the world. For a number of countries he was able to develop "pedigrees" showing the evolution of stem rust races by steps involving changes in single loci for pathogenicity. The data are particularly clear for Australia where the sexual cycle is of little importance. Occasionally a single new race became predominant and then evolved through single-gene mutations. For example, race 21 became widespread in eastern Australia in 1954 (Luig 1983) and then went through a long process of evolution (Fig. 2.7). It is interesting to note that one race may evolve from another by several different routes. For example, the race 15 phenotype evolved from race 21 by five different paths (Fig. 2.7). Also the same mutation can occur at several different times in the evolutionary process.

When new mutants for virulence have been observed in the field, they seem to be fully virulent. In the greenhouse, Watson and Luig (1968) inoculated Yalta (*Sr11*) with large quantities of spores of an avirulent, gray-brown culture of stem rust. The normal infection type on Yalta was a fleck (0;). A mutant that gave an X= infection

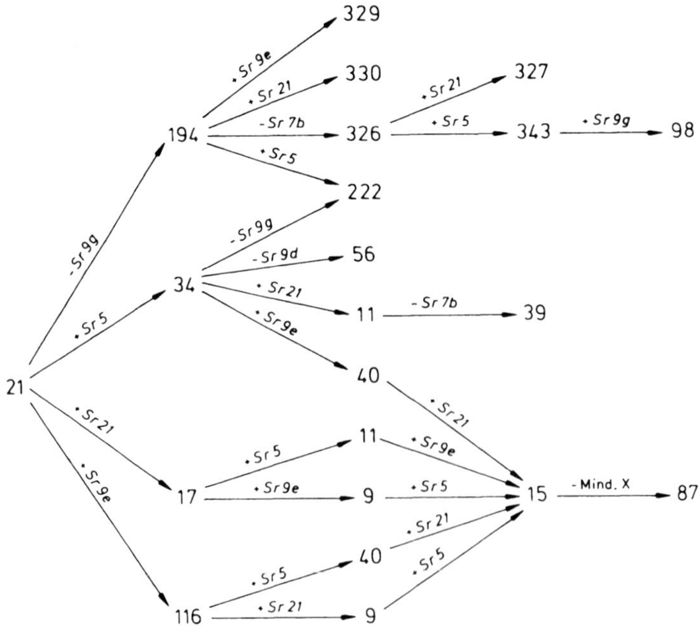

Fig. 2.7. Australian races of stem rust derived from race 21 as a starting point (a modification of Fig. 6 in Luig 1983). A *minus* indicates the loss of virulence on a gene for resistance and a *plus* indicates a gain in virulence

type (IT) was obtained and when the process was repeated, a mutant that gave a $2^{+}3$ IT was obtained from the first mutant. Similarly, they found that field cultures of stem rust could be classified into four types based on their virulence at different temperatures on lines carrying *Sr6*. McIntosh and Watson (1982) indicate that further studies have found similar progressive increases in virulence in both stem and leaf rust.

Back mutations from virulence to avirulence, i.e., from recessive to dominant, are probably much less likely to occur. Mutation is presumably a random process. The chance that a random change at a locus will result in the production of an active gene product (a dominant mutation) is very small. Furthermore, unless a recessive gene for virulence is deleterious, the dominant gene will have no competitive advantage and is unlikely to become frequent in the population. Nevertheless, Luig's (1983) charts on the evolution of rust races show a number of changes from virulence to avirulence. They are particularly frequent for *Sr7b, Sr9d, Sr9g,* and Kota, but also occurred at least once on *Sr5, Sr9e,* and Mind. X. Either these are backmutations or they resulted from some other process such as somatic hybridization (see next section).

The frequency of mutation in the field has not been studied in controlled experiments. However, from field race surveys it is evident that mutation rates vary greatly for different loci. Mutations from avirulence to virulence on *Sr5, Sr9e* and *Sr21* have occurred frequently in the field, while virulence on *Sr26* has never been found in Australia despite the fact that large acreages of cultivars carrying *Sr26* have been grown for many years (Luig 1979, 1983). Similarly, in induced mutation experiments using EMS, Luig (1979) found large variations in the frequency of mutations. Frequent mutations occurred for virulence on *Sr5, Sr9e,* and *Sr21,* but the field and induced mutation rates did not always correlate well.

Surprisingly, in a number of cases genes for virulence have arisen in a rust population although they were not necessary for its survival, i.e., there was no selection pressure from the growing of cultivars carrying the corresponding gene for resistance (Luig 1983). This can probably occur in two ways. The gene for virulence may have a selective advantage over the gene for avirulence, even in the absence of the corresponding gene for resistance in the commercial wheat crop. Since recessive genes are often deleterious, this is probably not a common situation. Alternatively, the gene for virulence may have been present in a race that became established and then spread by asexual reproduction because it was particularly aggressive or carried other desirable genes for virulence.

In most wheat-growing areas, mutation is probably the most important source of variation and evolution in the rust pathogens.

2.6.3 Somatic Hybridization

In many fungi, variation can result from the fusion of hyphae from different genotypes, followed by nuclear exchange and the reassortment of different nuclei in binucleate or multinucleate cells. On wheat plants, mycelial cells of the wheat rusts are primarily dikaryotic, but multinucleate cells have occasionally been reported (e.g., Nelson et al. 1955; Goddard 1976a).

In studying somatic hybridization in the rusts, it is essential to eliminate contamination as a possible source of variants. In addition to complete isolation of the experiments, genetic markers such as spore color are often used. Mutation is another possible source of variants but its frequency is too low to be an important factor in most experiments. Somatic hybridization has been studied in all three wheat rusts.

Nelson et al. (1955) observed the fusion of hyphae from different stem rust biotypes on a thin agar film. They also studied 114 mixtures of two to seven races that were avirulent on Khapli. The mixture of urediospores was either put directly on Khapli or increased on Little Club (susceptible) and transferred to Khapli. Four new biotypes were produced, one of which was more virulent on Khapli and several other cultivars than was either parent. It showed a high degree of instability and gradually lost virulence on Khapli. From 25 single-uredium isolates from type 4 infections, both parents and two new biotypes were obtained. Urediospores and hyphal cells of the virulent heterokaryon were frequently tri- or tetranucleate, particularly in early generations (a heterokaryon contains genetically different nuclei). The two new biotypes and parental types derived from the heterokaryon were typically binucleate. Nelson et al. (1955) concluded that the virulent heterokaryon resulted from the reassortment of nuclei from the parents and that it was unstable because of its trinucleate condition. The fact that Khapli was resistant to all North American races of stem rust eliminated any possibility of contamination. However, using Khapli also probably greatly reduced the chances of detecting somatic recombinants. Heterokaryons virulent on Khapli were produced twice, reducing the possibility that mutation was involved.

Following the work of Nelson et al. (1955), several studies of somatic hybridization were carried out on stem rust, often using gray-brown or orange (yellow) spore colors as markers. Nelson (1956) mixed a red-brown spored race 11 avirulent on Vernal with a gray-brown spored race 121 virulent on Vernal. One orange pustule was obtained on Vernal. It gave rise to 32 isolates, 30 intermediate in virulence to the two parents and two more virulent than either parent. Watson (1957) mixed a yellow-spored and a red-brown spored culture and obtained four new cultures that could not have arisen by either contamination or mutation. The parent cultures had been increased from single spores and had produced consistent reactions for eight generations. Later, Watson (1981) reported that at least 20 new types had been obtained from mixtures of the two original parents. Ellingboe (1961) studied six mixtures, each involving race 111 and one other race. Anywhere from 1 to 19 new types were obtained from a particular mixture, a total of 57 in all. When tested on 54 cultivars or lines, all 57 isolates were different. All were virulent on one or more cultivars or lines on which both parents were avirulent. One mixture produced isolates with five different reactions on one host. At least three genetic loci are required to explain such a result.

In leaf rust, Vakili and Caldwell (1957) mixed a red-spored race 2, avirulent on five cultivars, with a yellow-spored race 122 virulent on the five. One hundred and forty-nine red pustules and seven mixed red and yellow were obtained from inoculations on the five cultivars. Sixty-seven nonparental clones were identified. They fell into 16 previously described races and 17 new races. However, in similar experiments other workers obtained negative results (Barr et al. 1964; Bartos et al.

1969). It is not clear why conflicting results have been obtained with leaf rust or why the results differ from those with stem rust.

Somatic hybridization in yellow rust was first demonstrated by Little and Manners (1969), who produced two new races. Goddard (1976b), Taylor (1976) and Wright and Lennard (1980) also produced somatic recombinants in yellow rust. The frequency of recombinants was surprisingly high.

Some of the products of somatic hybridization can be explained on the basis of the exchange of whole nuclei between isolates, and their reassortment. In theory it should be possible for a single dikaryotic isolate to give rise to somatic recombinants. However, although fusion does occur between hyphae of a single isolate, no instability and production of somatic recombinants have been found. Apparently, some mechanism prevents the reassortment of nuclei and the production of homozygotes in such a situation.

When two dikaryotic isolates are mixed, four new combinations of nuclei are possible. Since virulence is usually recessive, genes for virulence in one nucleus can be masked by dominant genes for avirulence in the second nucleus of a dikaryon. For example, if two isolates have the following genotypes (the genotypes in brackets are the genetic constitution of the individual nuclei of a dikaryon):

(*P5P6*) (*p5p6*) – avirulent on *Sr5* and *Sr6*.
(*P5p6*) (*p5P6*) – avirulent on *Sr5* and *Sr6*,

then somatic recombination can result in four new dikaryons, assuming that the nuclei can recombine in all combinations regardless of mating type:

(*P5P6*) (*P5p6*) – avirulent on *Sr5* and *Sr6*.
(*P5P6*) (*p5P6*) – avirulent on *Sr5* and *Sr6*.
(*p5p6*) (*p5P6*) – avirulent on *Sr5*, virulent on *Sr6*.
(*p5p6*) (*p5P6*) – avirulent on *Sr5*, virulent on *Sr6*.

All four are genetically different than the parents and two are phenotypically different. Depending on the number of loci involved and the genetic constitution of the nuclei in the two parental dikaryons, various other combinations of parental and nonparental phenotypes are possible. All four new dikaryons may be phenotypically different from the parents but that is the maximum number possible. Presumably homozygous dikaryons are not produced, just as there is no evidence that they can be produced by somatic recombination within a single dikaryotic genotype.

In several experiments, more than four new phenotypes have been produced from somatic recombination involving two parents (e.g., Vakili and Caldwell 1957; Ellingboe, 1961; Watson 1981). These examples can be explained only by some mechanism other than simple reassortment of the whole nuclei. Since nuclei in a dikaryotic cell normally divide simultaneously, one possibility is that the two nuclei divide in the same plane and some chromosome are exchanged during mitosis. However, at least for yellow rust, Wright (1976) found that the nuclear envelopes remain intact during metaphase and chromosomal exchange could not take place.

A second possibility is that a parasexual cycle occurs as has been reported in a number of other fungi. As outlined by Pontecorvo (1956), a parasexual cycle involves five steps:

1. Fusion of unlike haploid nuclei in a heterokaryon.
2. Multiplication of the resulting diploid nucleus.

3. Eventual sorting out of a homokaryotic diploid nucleus.
4. Mitotic crossing over during the multiplication of the diploid nuclei.
5. Vegetative haploidization.

Steps 1, 4, and 5 occur very infrequently so the process is essentially impossible to study except by observing its consequences. A diploid stage has not been reported in the rust fungi but this may be because it occurs very rarely. However, Williams (1975) reported that a uninucleate strain of *P. graminis* f.sp. *tritici* established in axenic culture (i.e, grown on a defined medium) was diploid.

The observed results from somatic recombination experiments in the rusts could be explained by:

1. Hyphal fusion of two dikaryons followed by reassortment of nuclei to form new dikaryons.
2. Fusion of two nuclei in a cell to form a diploid nucleus.
3. Possibly one or more mitotic divisions during which mitotic crossing over could occur.
4. Vegetative haploidization (sometimes called somatic reduction) and reassortment of chromosomes.
5. Formation of new dikaryons.

Only step 1 is known to occur. Step 3 is not essential since vegetative haploidization will result in new combinations of chromosomes and genes not found in the parents (Fig. 2.8). However, if some genes for virulence are linked, mitotic crossing over will increase the number of potential new gene combinations.

Watson and Luig (1958) showed that somatic recombination could occur in the field. However, there is no way of determining whether it does occur in nature or, if

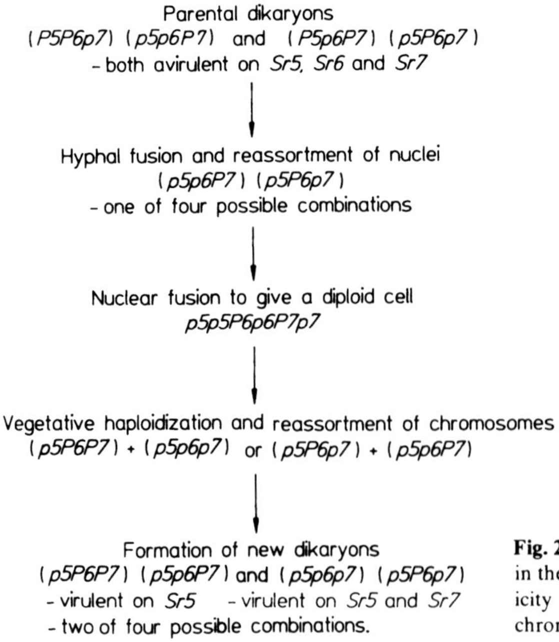

Fig. 2.8. Possible steps in a parasexual cycle in the wheat rusts. Each locus for pathogenicity is assumed to be on a different chromosome

it does, how frequent it is. Nevertheless, considering how often natural epidemics of the rusts involve mixtures of races, it would be surprising if somatic hybridization did not occur at least occasionally. The results reported by various authors indicate that somatic hybridization is much more likely to occur with some combinations of biotypes than with others.

2.7 Eradicating Barberry to Control Stem Rust

The common barberry (*Berberis vulgaris* L.) had a number of uses – medicinal purposes, handles for tools, hedges, fruit for jelly and wine, and a yellow dye from the bark – and was grown widely in Europe (Roelfs 1982). It was introduced to North America and spread widely. European farmers recognized the relationship between barberry and stem rust and in 1660 a law was passed in Rouen, France, prohibiting the growing of barberry (Roelfs 1982). Similar laws were passed in the eastern United States in the 1700s.

Beeson (1923) found that in an area of Indiana where yields of unrusted wheat were 15–25 bushels per acre, stem rust from a single barberry bush reduced yields to 0 in some nearby fields and to 10 bushels per acre in fields 2 miles away. This is not surprising considering that Stakman et al. (1927) estimated that a single average-sized barberry bush could produce 64 billion aeciospores.

In 1916, a stem rust epidemic in the main spring wheat-growing areas of the United States and Canada caused an estimated loss of 300 million bushels of wheat (Roelfs 1982). This stimulated an eradication program that began in 1918. In the great plains area of North America, by 1930 barberry was no longer important in the occurrence of stem rust epidemics and the frequency of epidemics has been greatly reduced. By 1967, almost 100 million barberry bushes had been destroyed (Roelfs 1982). As pointed out by Roelfs (1982), barberry eradication has four main effects: (1) Delaying the onset of rust which is then initiated only by urediospores spreading north from overwintering areas in the southern United States (2) reducing the initial inoculum level which often is heavy around barberry bushes but is light when it depends on airborne urediospores traveling long distances, (3) decreasing the number of pathogenic races by preventing the production of new races through sexual recombination, (4) stabilizing the pathogen population by reducing the number of races found each year and the frequency of changes.

CHAPTER 3

Rust Tests Under Controlled Conditions

3.1 Introduction

To study the pathology, physiology, and genetics of the wheat rusts, and to breed for resistance, it is essential to have efficient and reproducible methods of producing infection under controlled conditions. For many purposes, such as breeding for resistance, it is usually sufficient to produce infection on all of the plants in a test. The test must be adequate to accurately determine the infection types on different host genotypes. Usually that is fairly easy to do. For some other purposes, it is necessary to have a uniform infection frequency throughout a test. This is much more difficult to obtain.

3.2 Production and Storage of Inoculum

The first step in controlled rust tests is the production of pure inoculum of one or more rust genotypes.

3.2.1 Hosts for Rust Increase and Universal Suscepts

A rust isolate can be increased on any host genotype that is susceptible to it. Experience shows that some hosts are more productive spore producers than others. If increases are to be done on seedlings, then a host genotype that has large, upright, long-lived seedling leaves should be chosen. Growing the seedlings at cool temperatures and high light intensity will help to produce upright growth and make spore collection easier. Larger increases can be obtained by rusting adult plants when the flag leaf is fully developed, but, of course, the process takes longer than using seedlings. Often when increasing a rust isolate, a particular host cultivar or line can be used that will help to maintain the purity of the rust. For example, to increase a specific stem rust isolate, a host can be selected that is highly resistant to leaf rust and to many isolates of stem rust except the one being increased.

A universal suscept, i.e., a host genotype susceptible to all genotypes of the particular rust pathogen and, therefore, carrying no genes for resistance, can be a useful host. It is even more valuable as a susceptible parent in crosses for genetic studies and as a recurrent parent in backcross programs to produce near-isogenic lines carrying single genes for resistance. Attempts have been made to produce a universal suscept by crossing parents with complementary susceptibility and selecting for complete susceptibility. However, the resulting lines have always been found to have resistance to some genotypes of the pathogen. To select a universal suscept, it is necessary to have a rust genotype that has no genes for virulence, i.e., it is avirulent on all host genotypes except the universal suscept. Unfortunately, such genotypes have not been identified. In Australia, Watson and Luig (1963) developed a highly susceptible line, W2691, from the cross Little Club//Gabo*3/Charter. However, it proved to be resistant to some isolates of *P. graminis* f.sp. *secalis* that can attack a few wheat cultivars (Roelfs and McVey 1979). Knott (1981) produced a day-length insensitive, highly stem rust susceptible line, LMPG, from crosses involving Little Club, Marquis, Prelude, and Gabo. Day-length insensitivity is a desirable character for lines that are to be used as hosts or parents under conditions where day lengths are short. However, LMPG proved to be resistant to some stem rust genotypes in Australia (R.A. McIntosh personal communication).

In various areas of the world, wheat workers have selected susceptible host genotypes that are particularly suitable for their work. For example, in North American studies with stem rust, Chinese Spring, Baart, Little Club, McNair 701, and Marquis are some of the cultivars that have been used as susceptible hosts and parents. For leaf rust, Thatcher spring wheat and Wichita winter wheat have been used and for yellow rust, Michigan Amber, Lemhi, and a line of *Triticum dicoccum*.

3.2.2 Growing Plants and the Use of Maleic Hydrazide

Plants for rust increase can be grown under any conditions that are favorable for plant growth. Moderate temperatures and good light conditions are desirable. Either greenhouses or growth chambers are satisfactory. In greenhouses, supplementary lighting from fluorescent lamps should be used whenever the days are short or the light intensity is low. A 14–18-h day length is desirable. The more vigorous and healthy the plants are, the better the rust increase will be.

If seedlings are to be used, they can be grown thickly, 20 to 30 in a 15-cm diameter pot, particularly if they are treated with maleic hydrazide. Maleic hydrazide applied at the right stage suppresses growth after the second leaf but causes the first two leaves to be vigorous and erect, and to remain green longer than normal. The uredia sporulate particularly heavily. A maleic hydrazide solution is prepared using 100–400 mg per litre of water. The optimum strength varies depending on environmental conditions. Rowell (1984) reported that 100 ppm was better during the winter and 200 ppm during the summer. About 100 ml of the solution is applied to the soil surface of a 15-cm diameter pot when the seedling coleoptiles are 2–5 cm high.

If older plants are to be used for rust increase, they should be planted thinly, 3 or 4 to a 15-cm pot to allow vigorous growth.

3.2.3 Inoculation Procedures

Many procedures have been used to deposit urediospores on the plant parts to be infected. The procedure varies depending on the amount of material to be inoculated, the source of spores, and the equipment available.

For relatively small numbers of seedlings, some workers begin by rubbing the leaves lightly between moistened fingers. This removes some of the wax and small droplets of water are more likely to adhere to the leaves.

Sometimes the initial source of inoculum is a dried leaf containing one or more pustules. The seedlings to be inoculated should be sprayed lightly with water, perhaps containing a wetting agent such as Tween 20 (about one drop per litre). The spores are scraped from the dried leaf using a moistened scalpel which is then brushed lightly on the seedling leaves. Placing a drop of water on the pustule or pustules may help the pickup and transfer of the spores. A simpler but cruder procedure is to wipe moistened fingers over the infected leaf and then wipe them lightly over the leaves to be inoculated.

If the inoculum source is a pot of rusted plants, they can be brushed against or shaken lightly over the moistened plants to be inoculated.

When bulk spores are available, several inoculation methods can be used such as dusting or spraying. For many years a common procedure was to mix spores with talc in about a 1:4 ratio. The mixture was then dusted onto the plants to be inoculated. Simple dusters are easily made using a bottle, flask, or test tube with a cork or rubber stopper, a couple of pieces of glass tubing and a rubber or plastic squeeze bulb to create air pressure (Fig. 3.1).

Spores can also be suspended in water or a light, nonphytotoxic mineral oil (Rowell and Hayden 1965) and sprayed on the plants. Because of problems with the suspension of spores in water, the use of a mineral oil such as Mobilsol 100 or Soltrol 170 is now widespread. Various types of atomizers are available, some of which operate off an air line while others use pressure cans.

With stem rust, if only a limited amount of spores is available, a suspension of spores in water can be injected into the upper internodes of a few adult plants, using a hypodermic needle. Plants at the flag leaf stage are best. Once a few pustules have appeared, the rust can be spread further by rubbing moistened fingers over the pustules and then over the uninfected leaves and stems.

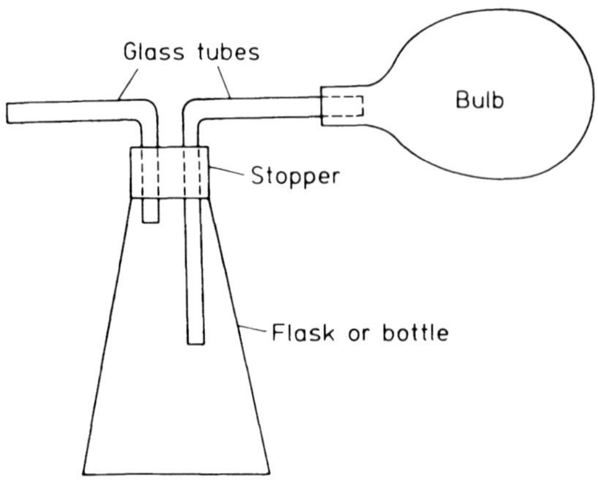

Fig. 3.1. A simple duster for inoculating wheat plants with mixtures of rust spores and talc

3.2.4 Environmental Conditions for Infection

The infection period is critical for the production of good rust infections. It must provide the conditions necessary for spore germination and the infection of the wheat plants. For all three rusts this involves a period in which there is free water on the inoculated plants. The temperature and light requirements are fairly similar for leaf and stem rust but much different for yellow rust.

The desirable infection conditions for leaf and stem rust are as follows:

1. A period with free water on inoculated plants at 15–24°C in the dark (Sharp et al. 1958). Commonly the plants are left in an infection chamber in the dark overnight at a temperature below 20°C. Cool temperatures make it easier to maintain high humidity and are favorable for spore germination, germ tube growth, and appresorium formation.

2. An additional period of 2 to 4 h with free water on the plants but at higher temperature (about 25°C) and in bright light, either artificial or sunlight. Higher temperature and bright light favor the development of the infection structures, particularly in stem rust.

3. A period of slow drying of the plants.

Clearly the favorable conditions for infection are similar to those that occur in the field – a cool night with dew deposition followed by rising temperatures and increasing light.

Yellow rust requires much cooler temperatures for good infection and development than do stem and leaf rust, and the effects of light and dark are less important. Providing suitable temperatures may require more elaborate equipment than is needed for leaf or stem rust, unless the work can be done during the winter when the weather is sufficiently cool. Temperatures of 10–15°C are suitable for the incubation period. Then about 15°C is best for good plant growth and rust development.

Many types of chambers have been used to hold plants during the infection period, some very simple and others very complicated. A simple procedure for leaf or stem rust is to inoculate the plants and place them in an open-topped, watertight container of a suitable size (or inoculate them right in the container). A large garbage can will do. The plants are then sprayed with a fine mist of water, preferably distilled. Otherwise the water should be checked to ensure that it does not inhibit germination. Some water is placed in the bottom of the container to increase the humidity. The top of the container is then covered with waterproof and light impermeable material – a sheet of plastic and then some cloth or canvas works well. If the temperature in the area around the container can be dropped, this will help maintain the humidity and the free water on the plants. Spraying the plants with water periodically is useful if the plants tend to dry off. The plants are usually left in the container overnight. Then the canvas cover is removed and the plants are illuminated with either natural or artificial light while still wet. If possible, the temperature should be raised to about 25°C. After 2–4 h the plastic is removed and the plants allowed to dry off slowly.

Some workers inoculate plants in small greenhouse compartments with built-in misting systems. Natural cooling and darkening can be used after inoculation and natural heat and light in the morning. Alternatively, artificial temperature and light controls can be employed.

For more critical work where reproducible environmental conditions are essential, elaborate dew chambers are used. They can be of various sizes and normally include systems for heating, cooling, and lighting. A pan at the bottom contains water which may be heated to increase evaporation. Inoculated plants are placed in the chamber and it is then cooled to, perhaps 15–18°C. This causes the deposition of dew on the plants. After 12–18 h in the dark, the lights are turned on and the temperature raised gradually to 25°C. After 2–4 h the plants are allowed to dry slowly and then returned to the growing area.

All of the equipment described above can be used for leaf or stem rust, and also for yellow rust if the necessary cool temperatures can be maintained.

3.2.5 Maintaining the Purity of Isolates

For many purposes it is essential to have pure inoculum of a single genotype of a single rust. Maintaining purity is not easy. During the growing season, if a crop is infected with one or more of the rusts, then urediospores will be present in the air and can easily get into most greenhouses. It may be unsafe to increase rust isolates during this period. Many workers handle a number of isolates of one rust or even of more than one rust. Every effort must be made to keep them separate and to prevent contamination. If possible, different isolates should be increased in different greenhouse compartments or growth chambers, or at different times. Inoculation should be done in an isolated area so that airborne urediospores are not carried to other plants. Fortunately, spores can produce infections only if appropriate environmental conditions, including free water, are provided. In addition, spores lose viability fairly quickly under normal light and temperature conditions. All equipment used in rust inoculation should be decontaminated before it is reused. Washing and then rinsing in 95% ethyl alcohol is satisfactory.

If a stem rust culture becomes contaminated with leaf rust, the leaf rust will often build up rapidly during successive increases. Apparently, a higher percentage of leaf rust spores are successful in producing infection than are stem rust spores. For the same reason it is easier to obtain consistent infections with leaf rust than with either stem or yellow rust. Contamination of a stem rust isolate with leaf rust can be prevented by increasing it on a host that is highly resistant to leaf rust. Alternatively, an application of the fungicide triazbutyl (Indar), which is effective against leaf rust but not stem rust, can be made to the soil surface of pots at least 2 days prior to inoculation (Rowell 1984).

3.2.6 Establishing Pure Isolates

If only mixed inoculum is available, the surest way to establish a pure isolate is by using a single spore to initiate an infection. Dry spores are dusted thinly on a glass slide which is put under a dissecting microscope at fairly low magnification (about

50x). A single, isolated spore is picked up on the tip of an animal hair that has been glued to a wooden handle or a glass tube drawn to a fine point. It is then transferred to the leaf of a wheat seedling near a spot that has been marked with India ink. The leaf should be checked under the dissecting microscope to be sure that only one spore is present. Since the infection rate with single spores is often low, a number of single spore isolations must be made. The inoculated seedlings are then put in an infection chamber under the appropriate conditions. At all stages the procedure must be carried out under rust-free conditions to prevent contamination. The inoculated plants are watched carefully for the appearance of flecks which indicate successful infections. Each infected plant must be then be placed in some form of isolation chamber. Spores from a single pustule are collected and increased further under isolation.

Many different types of isolation chambers have been developed, some quite simple, others more complicated. Often clear, cylindrical plastic chambers that will fit over or on single pots are used (e.g., see Browder 1971). Since light can cause damaging heat build-ups in transparent chambers, they must be well ventilated. Workers who make frequent use of isolation chambers usually provide for air flow using an air line.

Pure isolates can also be established using single pustules. The procedure is simpler but there is always a slight danger that what appears to be a single pustule may actually have arisen from two infections occurring at almost the same spot. To start the procedure, a few seedlings should be inoculated very thinly so that pustules that develop are well spread out. A young well-isolated pustule is selected and the piece of leaf containing it is cut out. Care must be taken to ensure that it is not contaminated with spores from nearby pustules. The pustule is then used to inoculate fresh seedlings either by brushing it against them or by scraping the spores off with a scalpel and transferring them to the new seedlings. The increase is carried out under isolation.

Rowell (1984) recommends selecting a seedling with a well-developed uredium near a leaf base. The leaf blade is cut off above the uredium and any other plants in the pot are removed. The plant and uredium are washed to remove spores and then placed in an infection chamber to induce germination of the remaining spores. The plant is kept under isolation for 48 h until a new crop of uncontaminated spores is produced and used for inoculation.

Some workers use detached leaf cultures for the increase of single spores. A number of leaf segments are floated in Petri dishes on a solution containing 1000 g of sucrose per litre and 50 parts per million of benzimidazole. A single spore is placed on each leaf segment. Each segment on which an infection appears is transferred to a separate Petri dish. When the pustule develops it is used to inoculate seedlings and increase the isolate.

If a single spore or single uredium is used to purify inoculum that has become mixed, it is essential to test the purified isolate to ensure that it is of the desired genotype and not an off-type or contaminant. This will involve testing the isolate on key host genotypes to confirm its identity.

3.2.7 Collecting Rust Urediospores

If small amounts of material are to be collected and stored for short periods, rusted leaves can be harvested and placed in glassine envelopes. Only a few leaves should be placed in an envelope and they should be dried rapidly at room temperature. This system is often used for field collections, particularly during survey work.

Two procedures are commonly used for collecting spores from infected seedlings or adult plants. A simple procedure is to hold the plants over a suitable collecting surface or container and tap them gently so that the spores fall to the surface. The collector may be a sheet of smooth paper or cardboard, wax paper or aluminum foil, a large glass funnel, etc. If the spores are on a flat surface they can then be gathered in one place by scraping them together with the edge of a smooth ruler or plastic pot label, or by tipping the collecting surface and forming a spout. A disadvantage of this procedure is that spores are released into the air and may contaminate nearby plants and equipment.

Cyclone collectors that use vacuum to suck up spores can also be employed. Various sizes can be made ranging from small ones that can be used to collect spores from a single pustule to large ones that can be used for mass collections (Tervet et al. 1951).

Whichever system is employed, great care must be taken to prevent contamination. Any equipment that is used should be washed and sterilized with 95% ethanol before being reused.

Spores collected by either method will often contain bits of plant material, soil particles, and small insects such as aphids. They can be cleaned out by putting the material through a fine sieve.

3.2.8 Storage of Urediospores

For good storage, urediospores should be dry. If the humidity is low, spores can be air dried at room temperature for about 24 h, but not in sunlight. However, if the natural humidity is high, the spores should be dried in open containers placed in a desiccator containing a desiccant such as calcium chloride, silica gel, or KOH (about 80 g/100 ml of water). Dry spores can be stored in sealed containers in a refrigerator at 2–5°C. Viability will decline gradually but normally some spores will germinate for at least 6 months and sometimes for a year. Leaf rust appears to survive better than stem rust.

For people working with many rust isolates a system of long-term storage is essential. Otherwise the regular increase of large numbers of isolates becomes a time- and space-consuming chore. Sharp and Smith (1957) developed a vacuum-drying method that extends the life of spores for several years. Spores in lyophil tubes are connected to a manifold and through a moisture trap to a high vacuum pump. The spores are dried for 2 or 3 h to about 2% moisture and the tubes are then sealed using a high temperature gas burner. Care must be taken to avoid heating and killing the spores. The life of the dried spores is lengthened considerably, even when they are held at room temperature. Sharp and Smith (1957) reported reasonable viability for up to 5 1/2 years, if the tubes were stored at 2–4°C. However, there was considerable variability in different batches of spores, probably depending on the conditions under which they were produced. Vacuum-dried spores germinate poorly unless they are rehydrated slowly before inoculation. Usually it is sufficient to open the tubes and leave them in a laboratory at room temperature for 24 h. Alternatively, the open tubes can be placed over water in a sealed container such as a desiccator for up to 24 h.

When liquid nitrogen became readily available, Loegering et al. (1961b) found that dry spores could be stored in heat-sealed glass tubes in liquid nitrogen for long periods. Liquid nitrogen evaporates slowly even from special cryostats, so a readily available source of liquid nitrogen is necessary to regularly refill the containers. This can be expensive. Glass tubes must be handled with care using gloves and a face shield. If the seal is not complete, liquid nitrogen will leak into the tube and it will explode when removed from the cryostat. This can be prevented by storing the spores in plastic, screwtop vials of the type used to store cattle semen. At the Plant Breeding Institute at the University of Sydney, spores are stored in special, foil packets about 5×5 cm that are heat-sealed. The packets take up much less space than tubes and are more easily handled. Rowell (1984) reports the use of larger polyethylene bags which, however, are more fragile. Spores taken directly from liquid nitrogen and thawed at room temperature germinate poorly, if at all. However, germinability can be restored by immersing the sealed tubes or packets in water at 40–45°C for 5 min, or by opening the tubes or packets and rehydrating the spores in a closed container over water for 16–24 h. Rowell (1984) reported good germination and infection with spores stored for 18 years. Spores remain viable after repeated cycles of freezing and thawing.

More recently it has been found that spores can be stored in ultralow-temperature freezers at about -50°C, at least for several years (Rowell 1984). The usual temperature shock or rehydration must be given when the spores are removed.

3.2.9 Testing Spore Germination

Spores should be tested for germination both before they are placed in storage, to ensure that they are viable, and when they are removed for use, to ensure that they have retained their viability. Although it should be simple, an accurate test of germination is not easy. Many factors can affect germination – temperature, light, water source, method of distributing spores on a surface, etc..

For most purposes, highly accurate germination counts are not necessary. It is sufficient to know whether germination is low, medium, or high so that inoculation rates can be adjusted. In this case it is usually satisfactory to put a thin layer of glass-distilled water in the bottom of a Petri plate and float a few spores on it. The number of spores should be limited because they produce inhibitors which can reduce germination. Tap water or even copper-distilled water may inhibit germination. The Petri plates should be kept in the dark at about 20° ± 3°C. Germination begins rapidly and a good idea of the rate can usually be obtained in 2 or 3 h. However, the percentage may increase for several more hours. The spores can be examined under a microscope at low power. If the layer of water in the Petri plate is not too deep, there will be less problem with spores floating out of the field of vision.

To prevent this problem, germination tests can be run by dusting spores lightly on the surface of 1% agar in Petri plates. Rowell (1984) found large differences in germination depending on whether spores were smeared on an agar surface (27%), dusted on (73%) or dispersed in a drop of light mineral oil and placed on the surface (99%).

For accurate determination of spore germination, Rowell (1984) outlines a procedure using 1% agar in double glass-distilled water. Spores are diluted in a 1:1 mixture of light inoculation oil and light petrolatum USP, and individual drops are placed on the agar surface. The inoculated plates are incubated in the dark at 18°C for 16 h and counts made. Since the spores are on agar, they do not float around and can be counted easily.

3.2.10 Other Foliar Diseases and Pollutants

If rust increases are done in greenhouses or growth chambers in which the humidity is high, powdery mildew (*Erysiphe graminis* DC) can become a serious problem. Other pathogens are less common. The old method of controlling mildew was to place powdered sulfur on steam pipes. The vapor controls powdery mildew but does not affect rust. The same effect can be produced by heating sulfur in containers on an electric hot plate. The temperature must be sufficient to cause vaporization but not oxidation of the sulfur since sulfur dioxide can injure plants. Newer fungicides are available that control mildew without affecting rusts. For example, 50 ml of a 40 ppm solution of systemic fungicide, ethirimol (Millstem), can be applied to a 10-cm pot at planting. It will control mildew for 4–5 weeks (Rowell 1984).

Air pollutants such as sulfur dioxide and gases from automobile exhausts are known to affect spore germination. They may occasionally cause problems in rust tests.

3.3 Seedling Rust Tests

Seedling rust tests for plant breeding purposes, genetic studies, etc., are carried out in basically the same way as the production of inoculum. The tests will probably be on a larger scale. Since the spores produced will not usually be harvested for inoculum, a small amount of contamination is of less concern. Nevertheless, it is desirable to be as careful as possible to avoid contamination.

Although some elaborate chambers have been developed for rust testing, it is quite possible to obtain satisfactory results with relatively simple equipment. The basic requirements are a duster or sprayer for inoculation and a system for maintaining free water on plants for a number of hours.

At the University of Saskatchewan, seedlings for rusting with leaf or stem are often grown in a greenhouse directly in soil beds, 1.2 × 2.4 m, or in pots placed on the beds. The beds have pipes standing in each corner to which a 1.2 × 2.4-m wooden frame covered with wire can be attached. The plants are inoculated in the late afternoon and the frame is covered first with clear plastic and then with canvas to make an infection chamber. A humidifier is placed in the chamber and run continuously. The temperature is allowed to drop to about 15°C overnight. In the morning the canvas is removed, the lights are turned on and the temperature raised to 20–25°C. After about 2 h the humidifier is turned off, the sides of the plastic are raised, and the plants allowed to dry off slowly.

In seedling tests the infection types are classified according to the appropriate system as described in Sect. 2.3.

3.4 Adult Plant Rust Tests

Some genes for rust resistance are effective only in adult plants, e.g., *Sr2, Lr12, Lr13, Lr22a, Lr22b, Yr11, Yr12, Yr13, Yr14,* and *Yr16*. Adult plants can be grown and inoculated under controlled conditions but it is more difficult to get uniform infection and to make accurate readings than is the case for seedlings. Pustule size can vary considerably on different parts of the plant, probably depending partly on the age of the tissue. Adult plant resistance may be expressed differently in the greenhouse than in the field. For example, the Canadian cultivar Manitou has resistance to some races of leaf rust in the field but adult plants appears to be susceptible to the same races in the greenhouse. In the field, plants carrying *Sr2* show good resistance to many races of stem rust, which is expressed as a reduction in pustule size and number, with the peduncle being almost free of rust. However, the author's experience has been that it is almost impossible to detect the effect of *Sr2* in adult plants in the greenhouse.

3.5 Tests with Accurate Control of Inoculation Rates

For some purposes it is necessary to have uniform inoculation frequencies throughout an experiment. For example, some types of resistance involve a reduction in the frequency of successful infections. To identify and measure such resistance, it is essential to have reproducible methods of depositing uniform amounts of inoculum throughout a test and of producing uniform environmental conditions for infection. Other experiments may involve comparisons among treatments and accurate and reproducible methods are essential.

Various types of settling towers have been developed to provide uniform deposition of urediospores on wheat plants (reviewed by Rowell 1984). Most towers consist of a cylindrical chamber. The bottom includes a turntable on which pots of plants are placed, and which turns slowly as the inoculation is carried out. In some cases, each pot is placed on a separate plate which turns slowly in the opposite direction to the main turntable. Either dry spores or spores suspended in light mineral oil are injected into the upper part of the chamber and allowed to settle on the plants below. Spores can be injected using a modified CO_2 pistol for dry spores or an atomizer for spores in oil. There must be a uniform cloud of spores in the upper part of the chamber. The rotation of the plants as the spores settle should provide uniform spore deposition. However, the rate of deposition can vary greatly depending on the attitude of the leaves, something that varies with different genotypes. Usually spores are deposited heavily on only one side of a leaf blade. To increase uniformity of deposition, some workers fasten leaves to a horizontal surface.

When the plants have been inoculated, they are moved to an infection chamber with accurate controls on temperature and light. Either dew formation or a humidifier can provide the free water on the leaves that is necessary for urediospore germination. The plants are then transferred to a chamber with accurate light and temperature control to complete rust development.

In experiments with controlled spore deposition, in addition to reading infection types, accurate counts of the number of uredia per square centimeter are normally made. Area meters can be used to measure leaf areas. If such equipment is not available, the area for the main part of the leaf in which the leaf edges are roughly parallel can be approximated by multiplying the length by the average width. If the infection is heavy, counting uredia becomes very time-consuming. An alternative procedure is to estimate disease severity using the modified Cobb scales of Peterson et al. (1948) (Fig. 3.2). The infected leaves are compared with theoretical diagrams showing the frequency of uredia for particular percentage disease severities. Even at

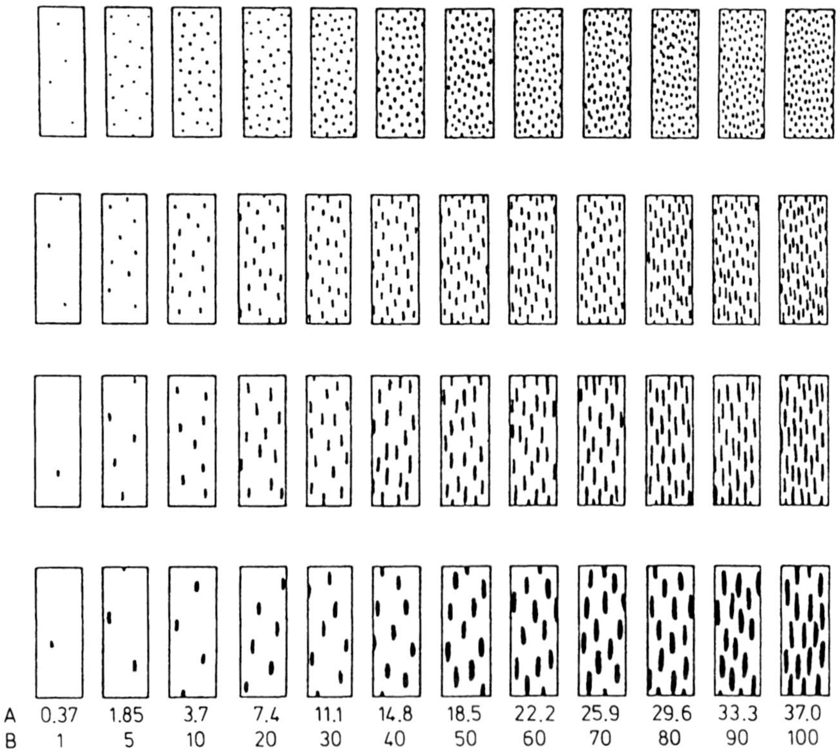

Fig. 3.2. The modified Cobb scale developed by Peterson et al. (1948). Used with the permission of Dr. A.B. Campbell

maximum levels of infection, only about one-third of the leaf or stem surface is covered with uredia. The diagrams take this into account – one third of the leaf area covered with uredia is considered to be 100% infection. For stem rust Rowell (1984) considers ten uredia per culm to be approximately 1% severity.

For some purposes it may be desirable to measure the length of the latent period, i.e., the period between inoculation and the eruption of pustules. Two genotypes may have similar infection types but differ in the length of the latent period. A long latent period in the field can slow a rust epidemic and provide a considerable degree of resistance. Since the length of the latent period is affected by environment, it should be measured only when plants are grown under controlled light and temperature conditions.

3.6 Field Rust Tests

Field rust tests are particularly important in wheat breeding and are fairly simple to carry out. They are described in detail in Chap. 8.

CHAPTER 4
Surveying Variability in Host and Pathogen

4.1 Race Surveys in the Rusts

4.1.1 The Purpose of Race Surveys

Since the discovery of pathogenic variability within the formae speciales of each of the three wheat rusts, race surveys have been carried out in the major wheat growing areas of the world. Initially their major purpose was descriptive – to determine the range of variability within each of the wheat rusts. In the process they provided pure isolates for use in a variety of studies, genetics, breeding, pathology, and physiology.

As breeding for rust resistance developed, race surveys provided essential information to determine the direction of breeding programs. A major objective was to detect new, highly virulent pathogen phenotypes as they appeared and before they became a threat to the crop. Knowledge of the occurrence of either new genes for virulence or new combinations of genes was essential for wheat breeders in determining the resistance necessary in breeding programs. Potentially dangerous races could be used both to evaluate potential parents and to select for resistance in the progeny of hybrids.

In addition to providing information on new rust genotypes, race surveys track changes in the frequencies of races and of virulence on specific resistance genes. Thus, they provided forewarning of the breakdown of resistance in commercial cultivars. For example, race 15B of stem rust was first detected in North America in 1939 and became predominant in 1950. It attacked most commercial wheat cultivars and resulted in a major epidemic in 1954. However, once resistant cultivars were produced, particularly in the spring wheats, the stem rust population stabilized. For a period the old race 56 became important again and since then the race 15 group has been predominant. No major new threat has developed.

Race surveys are also important in epidemiological studies. Since races are not distinguishable morphologically, pathogenicity is used as a marker to study the movement of spores of different races. For example, in recent years the race 15 group of stem rust has predominated in the central and northern Great Plains of North America. However, it has not been found in southern Texas and Mexico and, therefore, could not have overwintered there. Roelfs (1984) concluded that it overwintered in northern Texas and southern Oklahoma. Spectacular examples of spore movement have been detected in Australia (Luig 1985). On three occasions races of stem rust with completely new combinations of virulence have appeared and it was concluded that they originated in Africa and were blown across the Indian Ocean to Western Australia.

4.1.2 Sampling Rust Populations

The objective of the sampling determines how the collections are made. If the objective is to determine the composition of the rust population in an area, then some form of random sampling is essential. Unfortunately, that can be very difficult. Most rust will be found in commercial fields, but genes for resistance carried by the cultivars will determine the races found on them and samples taken from them will not be random. If cultivars susceptible to most races are still widely grown, collections can be made from them. Sometimes samples can be collected from wild species such as *Hordeum jubatum,* which is susceptible to stem rust. Nevertheless, it is difficult to be sure that random samples of the spore population have been obtained. Samples should never be taken near inoculated nurseries.

Eyal et al. (1973) used mobile nurseries to sample spore populations of powdery mildew (*Erysiphe graminis hordei*) on wild barley (*Hordeum spontaneum*) in Israel. Seedlings of a set of near-isogenic lines and accessions with known resistance were grown in trays. The trays were taken to various locations and the seedlings exposed for anywhere from 30 min to 72 h. They were then brought back to the laboratory, incubated, and the infection types recorded. In this way potentially dangerous strains of powdery mildew were discovered. In England the procedure has been modified by exposing the seedlings on top of a moving vehicle (Wolfe et al. 1981a). A similar procedure does not appear to have been tried for the rusts.

In most cases a survey is done by intensive sampling in cultivated fields. A good example is the rust survey carried out by the Cereal Rust Laboratory in the U.S.A. (Roelfs 1984). During the growing season about six trips of 25,000 km are made with stops at the first small-grain field after every 32 km (the survey covers all rusts on the small grains). The trips are made through major small-grain producing areas where rust can be a problem, and start in the south early in the season and move farther north as the season progresses. Collections taken from commercial fields or wild hosts by cooperators are also included. Large numbers of samples are taken for the major rusts and provide the basis for data on the frequency and distribution of races. These are published annually.

Extensive surveys take considerable time, labor, and money. Often the most important information to the wheat breeder is data on the occurrence and frequency of new genes for virulence and on the occurrence of virulence on resistant commercial cultivars or lines with particular gene combinations. Valuable information can be obtained by careful observation of and collections from fields of known cultivars. This can be supplemented by planting trap nurseries containing lines of known genotypes, including lines that are resistant to the prevalent races. Specific host genotypes will be attacked only by rust genotypes carrying the corresponding genes for virulence, and this will demonstrate their presence. Data from collections made in trap nurseries or from specific cultivars should not be included in regular survey data. Otherwise the survey would be biased in favor of rust genotypes virulent on specific hosts.

An alternative way of testing for new or unusual virulences is to composite urediospores from many locations and inoculate known host genotypes as suggested by Browder (1971).

4.1.3 Identifying Rust Collections

The use of the standard sets of differential hosts and of supplementary hosts has been discussed in chapter 2. In setting up a race survey, wheat breeders and pathologists need to work together to determine their objectives and, therefore, the differential hosts to use. If it is important to be able to compare the results with data from other parts of the world or with previous data from the area, it may be necessary to use some or all of the standard differential set. However, increasing use is being made of differentials carrying single resistance genes and particularly sets of near-isogenic lines when they are available. At the moment they are primarily available for stem and leaf rust. Production of near-isogenic lines for resistance to yellow rust has been more difficult (R. Johnson personal communication). The main advantage of using single gene differentials is that they provide information about the actual frequency in the rust population of virulence for specific genes. The wheat breeder will particularly want to test for new and potentially dangerous combinations of virulence in the rust. This may require the testing of currently resistant commercial cultivars and of resistant parents for the breeding program.

Usually samples from surveys consist of individual leaves from plants. Even pustules on a single leaf will often be of different genotypes if the rust population is heterogeneous. If a bulk collection of spores is used to inoculate a set of differential hosts, a range of infection types will almost certainly occur on some hosts. Thus, it may be impossible to classify the infection type on a host or it may be classified incorrectly as a type X. For this reason, most pathologists first inoculate each collection onto a susceptible cultivar. A number of single uredium isolates, often three or four per collection, are then made. A single stem rust uredium will normally produce enough spores to inoculate a differential set, particularly if the host plant has been treated with maleic hydrazide (see Sect. 3.2.2). Because of the small size of leaf rust uredia, one generation of increase is usually necessary before enough spores are available to inoculate a differential set.

Once pure isolates are available, the final step is to test them on an appropriate differential set of hosts. Where many isolates are being tested, mechanized systems of seeding large numbers of pots with the differential hosts have been developed. In a system described by Browder (1971), containers are filled with soil to a uniform level, a punching head is used to punch the required number of holes in the soil and a seeder then drops the required number of seeds of each differential in the holes. The seeder can plant 1, 5, 9, 16, or 30 lines in a container depending on its size. A complete set of differentials can be planted in one container. Although the lines are planted very close together, the use of punched holes keeps them clearly separate from one another. Mechanical systems speed up the process but similar procedures can be done manually.

Seedlings of the differential set are inoculated at the two-leaf stage. Care must be taken to ensure that each isolate is kept pure and there is no contamination during inoculation. Since environment, particularly temperature, can affect infection types, the seedlings must be incubated under controlled conditions, preferably in a growth chamber. A temperature of 18–20°C and good light is desirable for leaf and stem rust. Yellow rust requires cooler temperatures, preferably 15°C or lower.

When the uredia are fully developed, usually about 12 to 14 days after inoculation, the infection types are read using one of the systems described in Chapter 2. The infection types are used to determine the race classifications of the isolates. Stakman and Levine (1922) classified infection types of stem rust into three categories, resistant (0 to 2), mesothetic (X) and susceptible (3 and 4), for the purpose of race identification. Races were assigned numbers in chronological order as they were identified. Early workers with leaf and yellow rust followed a similar system. However, there were occasionally difficulties in classifying intermediate infection types and also a trichotomous key resulted in an enormous number of potential races. For stem rust, with 12 differentials there are 3^{12} or 531,411 potential races. Most recent systems use a dichotomous key with infection types classified simply as low or high, and thus reduce the potential number of races.

4.2 Surveying Resistance in the Hosts

4.2.1 The International Virulence Gene Survey

A major problem in comparing and integrating the rust work in various countries of the world is that both the rust populations and the host populations differ from country to country. Partly to overcome this problem, the First International Congress of Plant Pathology held in London in 1968 recommended that surveys be done for a number of fungal pathogens to determine differences in genes for virulence in various regions throughout the world. Dr. I.A. Watson of the University of Sydney agreed to carry out a survey for stem rust and the results have been reported by Luig (1983). The objectives of the survey were:

"1. to identify genes which condition resistance in all countries;

2. to assess the value of individual known and unknown genes for breeding programs in each cooperating country;

3. to predict the effectiveness of particular gene combinations;

4. to identify genotypes which possess new or unknown resistance genes;

5. to determine on a world-wide basis the distribution of single virulence genes and of virulence gene combinations;

6. to trace major rust migrations which have occurred in the past; and

7. to gain an understanding of the evolution which has taken place in regional populations of *P. graminis tritici.*"

To carry out the survey, sets of 117 selected lines and cultivars were sent to 18 cooperating institutions throughout the world. The set included 74 lines or cultivars with single genes, known gene combinations or combinations including unknown genes; 39 lines or cultivars known to be highly resistant; and 4 susceptible lines or cultivars (Line E W3498, W2692, Marquis, and Chinese Spring). The International or Stakman differentials were not included but it was assumed that seed was available at each institution. The cooperators carried out seedling tests with from 2 to 66 isolates from their countries. Luig (1983) summarized the results for each of the

117 entries. The data provided a wealth of information about the usefulness of individual genes and gene combinations in various areas of the world and about the evolution of virulence. It also identified rust isolates that carry unusual genes for virulence and combinations of genes and are, therefore, valuable testers. Luig (1983) also compiled a useful list of a large number of wheat cultivars and the genes for stem rust resistance that they are known to carry.

Luig (1983) did an extensive survey of the literature on stem rust in the major wheat-growing areas of the world. For a number of countries he was able to show the genetic relationships among the races of stem rust and their probable evolutionary origin.

4.2.2 International Testing of Wheat Germplasm

Over the years various organizations have been involved in international testing of wheat germplasm for rust resistance. For many years, the U.S.D.A. coordinated several tests. Now the testing is mainly coordinated by the International Agricultural Research Centers, particularly CIMMYT (the International Maize and Wheat Improvement Center in Mexico) for bread wheat and ICARDA (International Center for Agricultural Research in Dry Areas, Aleppo, Syria) for durum wheat. Each year CIMMYT sends out a large number of sets (186 in 1986–87) of the International Bread Wheat Screening Nursery containing about 250 entries and of the International Disease Trap Nursery of about 200 entries (245 sets in 1986–87) (Rajaram et al. 1988). These are grown throughout the world and readings on one or more of the rusts are taken at many locations. The results of the tests provide excellent data on the resistance of potential cultivars and parents to the three rusts. ICARDA distributes seed of a number of nurseries for durum wheat and again good data on rust resistance are obtained. An international Winter Wheat Performance Nursery of about 30 cultivars is coordinated by the U.S. Department of Agriculture and the University of Nebraska and is distributed to about 100 locations each year. Data are obtained on all three rusts.

A major objective of the CIMMYT wheat-breeding program is the development of germplasm for use by other breeders in their national programs. In breeding for resistance to the rusts, the CIMMYT wheat breeders make use of resistant lines collected from all over the world. Particular use is made of "hot spots", locations where there is maximum variability of a pathogen or maximum severity of a disease. Currently the locations used are Njoro, Kenya for stem rust, Quito, Ecuador for yellow rust, and Cd. Obregon and Rio Bravo, Mexico for leaf rust (Rajaram et al. 1988).

The various international nurseries provide very valuable data for wheat breeders in selecting rust-resistant parents to use in breeding programs.

4.3 The Analysis of Rust Populations

As more data were obtained on both pathogenicity in the rusts and resistance in wheat, it became evident that simple race identification did not provide all of the information that wheat breeders needed. Breeders are interested in the frequency of specific genes for virulence and of gene combinations, and changes in their frequencies. In particular they want to have early warning when new genes for virulence appear in a rust population and when rust genotypes develop that can attack previously resistant cultivars.

This change in emphasis led to an increased use of the methods of population genetics in analyzing populations of rusts and other plant pathogens. In using such methods, the basic assumption is that the rust population is discrete and that random genetic recombination occurs (Wolfe and Knott 1982). However, there is no known sexual stage in yellow rust, and the sexual stage is of little importance in many populations of leaf and stem rust, including those in Australia and the Great Plains of North America. Thus, many rust populations are mixtures of a relatively small number of asexually reproduced clones. As a result of selection, they are likely to be the clones that are most aggressive and best adapted to the host wheat populations. In addition, in carrying out rust surveys, samples are often collected over a large area and may not be from a single, discrete rust population. Analyses that combine data from different areas and different years in which gene frequencies may well be different, are subject to serious bias (Wolfe and Knott 1982).

Wolfe and Schwarzbach (1975) and Wolfe et al. (1976) proposed the analysis of the frequency of combinations of genes for virulence. If association is strictly random, then the frequency of a combination of two virulences should be the product of their individual frequencies. For example, if the frequency of virulence on A is 0.4 and on B is 0.2, then the equilibrium frequency of the combined virulences should be $(0.4)(0.2) = 0.08$. If the combination is not in equilibrium, i.e., the observed frequency is not equal to the expected value, then the action of some force such as selection is indicated. Wolfe and Schwarzbach (1975) showed that two populations of wheat powdery mildew in Sweden and the United Kingdom were in equilibrium. The sexual stage is common in powdery mildew.

Several authors have analyzed the association of virulences in the wheat rusts. Browder and Eversmeyer (1977) collected samples of leaf rust from eight geographic regions of the U.S.A. and tested them on nine wheat lines carrying single genes for resistance. For each region the frequency of virulence on each gene and each pair of genes was calculated. They then defined two coefficients:

1. The Virulence Association Coefficient (VAC) which is the frequency of the combined virulence on two genes, i.e., V1V2.
2. The Pathogenicity Association Coefficient (PAC) which is the frequency of the combined virulence on two genes plus the frequency of the combined avirulence on the two genes, i.e., V1V2 + A1A2.

Browder and Eversmeyer (1977) suggested that pairs of resistance genes for which PAC is large but VAC is small would be desirable combinations to use in controlling leaf rust because the data suggest that there is selection against the combined virulences. However, any pair of virulences in which one or both have low frequencies and neither has a very high frequency, will have small VACs and relatively high PACs, even if the virulences are in genetic equilibrium (Table 4.1). Thus, a low VAC is not necessarily an indication of selection against a particular combination of virulences. In addition, the data are probably biased by the fact that the populations were not the result of random mating.

Table 4.1. Changes in PAC and VAC with changing frequencies of virulence on *LrX* and *LrY*, assuming that the genes for virulence are in equilibrium

Frequency of virulence on		Value of	
LrX	*LrY*	PAC	VAC
0.70	0.04	0.316	0.028
0.50	0.04	0.500	0.020
0.30	0.04	0.684	0.012
0.10	0.04	0.868	0.004
0.05	0.04	0.914	0.002

For stem rust, Roelfs and Groth (1980) studied a large number of isolates from an asexually reproducing population from east of the Rocky Mountains in the U.S.A. where barberry has largely been eliminated, and from a sexually reproducing population. The latter were obtained from an area in Washington State where stem rust infection on wheat was initiated by aeciospores produced by the sexual cycle on barberry. The isolates were tested on lines carrying 16 known genes for resistance and the virulence frequencies determined for each population. It was assumed that virulence on each gene for resistance was controlled by a single gene. As would be expected, the sexual population was much more variable than the asexual population. In the sexual population, 100 virulence phenotypes were identified in 426 isolates compared to 17 phenotypes in 2377 isolates from the asexual population. The number of genes for virulence in the sexual isolates had a unimodal distribution with a range from 2 to 10 and a mean of about 6. The number of genes for virulence in the asexual isolates had a bimodal distribution with a low peak at 6 and a high peak at 10, a range from 5 to 12 and a mean of about 10. The difference in the average frequency of virulence genes in the two populations could have resulted from differences in the number of resistance genes carried by the hosts in the areas where the two populations were collected. Roelfs and Groth (1980) also showed that the number of distinct virulence phenotypes detected in the stem rust survey in the United States declined sharply with the eradication of barberry and the elimination of the sexual stage. For successive ten-year periods starting in 1918, the average number of phenotypes collected per year was 17.5, 10.7, 6.5, 7.7, 7.3, and 4.9.

Roelfs and Groth (1980) analyzed the observed and expected frequencies of all pairs of virulence genes. As expected the deviations between the observed and expected frequencies were much larger for the asexual than for the sexual population (typical examples for one set of 14 pairs of virulence genes are given in Table 4.2). As pointed out by Roelfs and Groth (1980), large deviations should be

Table 4.2. A comparison of the frequency of 14 pairs of virulence genes in asexual and sexual populations of stem rust in the U.S.A. (Data from Roelfs and Groth 1980)

	Frequency of virulence (%) on *Sr9a* plus													
	Sr5	*Sr6*	*Sr7b*	*Sr8*	*Sr9b*	*Sr9d*	*Sr9e*	*Sr10*	*Sr11*	*Sr15*	*Sr16*	*Sr17*	*SrTt-1*	*SrTmp*
Asexual population														
Observed	14	2	11	12	2	14	11	12	11	3	14	3	11	11
Expected	21	2	4	5	4	14	[a]	2	2	14	13	10	4	0
Difference	−7	0	+7	+7	−2	0	+11	+10	+9	−11	+1	−7	−7	+11
Sexual population														
Observed	44	[a]	12	14	1	44	2	61	2	59	40	47	12	2
Expected	47	[a]	7	11	[a]	44	[a]	57	[a]	60	41	50	10	0
Difference	−3	0	+5	+3	+1	0	+2	+4	+2	−1	−1	−3	+2	+2

[a] = less than 0.6%

expected in the asexual population because of the lack of recombination. In fact one virulence phenotype made up 69 percent of the asexual population and the 17 phenotypes fell into just six genetically similar groups. The genes in an asexually reproducing clone cannot recombine with genes in other clones except through rare somatic recombination and are effectively linked. Thus, the calculation of the expected frequency of pairs of virulence genes on the assumption of random recombination is not valid.

Extreme care must be taken in analyzing rust populations to ensure that the methods used are valid.

CHAPTER 5

Genetic Analysis of Resistance

5.1 Introduction

Biffen (1905), working with yellow rust, provided the first evidence that resistance to a pathogen could be controlled by a single, Mendelian gene. He showed that resistance to yellow rust in Rivet wheat was controlled by one recessive gene.

The occurrence of different races of a rust caused considerable confusion and disagreement in the early rust work. A cultivar that was resistant in one location sometimes was susceptible in another, and the stability of the rust pathogens was questioned. In particular, Ward (1903) developed a "bridging-host" hypothesis. He believed that when a rust was virulent on one host species but avirulent on a second, virulence on the second could be developed by culturing the rust on a species taxonomically intermediate between the two. The arguments were resolved with the work of Stakman and his coworkers, who tested a range of host genotypes and pathogen cultures (Stakman 1914; Stakman et al. 1918; Stakman et al. 1919; Stakman and Piemeisel 1917). They demonstrated the occurrence of different phenotypes for resistance within a host species and different phenotypes for pathogenicity within rust formae speciales (i.e., physiologic races). Perhaps most important, they showed that a physiologic race was a stable entity that produced consistent infection types and that resistance in the host was stable against specific races. This placed studies on the genetics of rust resistance on a sound basis.

5.2 Early Genetic Studies on Resistance

Much of the early genetic work on the inheritance of rust resistance in wheat was done in conjunction with wheat breeding. A susceptible cultivar would be crossed with a resistant parent. In the usual pedigree breeding system, F_2 populations and F_3 families would be studied, often in the field with natural rust epidemics. Since a separation of F_2 or F_3 plants into distinct classes was usually difficult, the analyses were frequently based on F_3 families. The families were classified into homozygous resistant, segregating, and homozygous susceptible classes. Sometimes segregating families could be classified into more than one type. In addition, in some cases seedling tests in the greenhouse with specific rust races were carried out.

As a result of these studies a considerable body of data on resistance to leaf and stem rust developed, which was summarized by Ausemus et al. (1946). Resistance was variously reported to be monogenic, digenic, or more complicated in inheritance. Ausemus et al. (1946) designated three genes for leaf rust resistance as *Lr1*-*Lr3*, and five genes for stem rust resistance as *Sr1*-*Sr5*.

Since these early studies were done in isolation, different workers using different wheat cultivars and different rust races, it was largely impossible to relate one study to another. Furthermore, many of them involved field studies with unknown mixtures of races, and results could vary from year to year. The results were complicated by the fact that a gene might provide resistance to one race and not to others. Two or more genes were often necessary to provide resistance to all races and the inheritance appeared more complex than it really was.

5.3 Procedures for Determining the Inheritance of Resistance to Rust

The rapid spread of race 15B of stem rust in North America in the early 1950's resulted in all commercial cultivars suddenly becoming susceptible. This greatly stimulated work on the inheritance of resistance. At the University of Saskatchewan, Knott and Anderson (1956) made use of two procedures to increase the efficiency of genetic studies: (1) diallel crosses among sets of resistant parents, (2) backcrosses of resistant parents to a susceptible line (i.e., testcrosses). Diallel crosses made it possible to determine whether resistant parents carried genes at the same or different loci. Backcrosses greatly simplified the genetic analysis as compared to the use of F_3 families.

When resistance is dominant, clearly identifiable, and simply inherited, its inheritance may be determined directly from the segregation of F_1 plants from a backcross. Usually, however, it is desirable or even essential to test F_2 families from the backcross. To determine the number of genes for resistance it is necessary to decide only whether each family is segregating or susceptible. If there is one gene for resistance, the ratio will be 1 segregating: 1 susceptible family, for two genes 3:1, for three genes 7:1, etc. The segregations within the segregating families often provide additional information on the types of gene action involved.

When a single gene is involved, the infection type on the F_1 and backcross F_1 plants will indicate whether the resistance is dominant or recessive. The conclusion can be confirmed from results in the segregating F_2 families from the backcross. Particularly when the gene conditions a moderately resistant infection type, it is not uncommon to find that the F_1 plants are intermediate in reaction and it may not be possible to classify plants within a segregating backcross F_2 family into distinct classes. If the classification is not clear, the plants should not be arbitrarily divided into resistant and susceptible classes by, for example, calling ITs of 2 and lower resistant and of 3 and above susceptible (IT = infection type). To be valid, a phenotypic classification must accurately reflect the genotypic classes and such a division often does not.

When two genes are involved, 3/4 of the backcross F_2 families will segregate and 1/4 will be susceptible, unless the two genes are linked. The segregating families will be of three types depending on whether they are segregating for the first gene, the second gene, or both genes. Whether the three types of families can be distinguished depends on whether the genes are dominant or recessive and whether they condition clearly distinguishable infection types (Table 5.1). Some of the possibilities are shown in the table, but the situation becomes more complicated if the genes show incomplete dominance.

Table 5.1. The various ratios that may be obtained when a resistant parent carrying two genes for resistance is backcrossed to a susceptible parent, the results depending on the dominance of the two genes and the infection types conditioned by them. The one quarter of the backcross F_2 families that are susceptible are omitted

Gene action	Infection Types	3/4 segregating families: 1/4 seg. for the first gene	1/4 seg. for the second gene	1/4 seg. for both genes
Both dominant	R, MR[a]	3R:1S	3MR:1S	12R:3MR:1S
		(may be distinguishable)		
	R, R	3R:1S	3R:1S	15R:1S
		(indistinguishable)		
One dominant and one recessive	R, MR	3R:1S	1MR:3S	12R:1MR:3S
	R, R	3R:1S	1R:3S	13R:3S
Both recessive	R, MR	1R:3S	1MR:3S	4R:3MR:9S
		(may be distinguishable)		
	R, R	1R:3S	1R:3S	7R:9S
		(indistinguishable)		

[a] R = resistant, MR = moderately resistant, S = susceptible. A resistant infection type might be a 0; or 1 and a moderately resistant a type 2.

When three independent genes are involved, there are seven types of segregating backcross F_2 families, three types segregating for one gene, three segregating for two genes, and one segregating for all three genes. It is unlikely that all seven types can be identified. Nevertheless, the ratio of segregating to susceptible families and the observed segregation within families will usually give a clear indication of the number of genes involved and the types of gene action.

In any testcross, it is essential to test enough families to clearly distinguish between the various possible ratios. The number required gets larger as the number of genes increases and, therefore, the ratio increases. To distinguish between a 1:1 and a 3:1 ratio, about 44 families are needed and to distinguish between a 3:1 and 7:1, about 110 families (Steel and Torrie 1980). Fortunately, the additional information provided by infection types and their segregation within families can help to determine the number of genes involved. A reasonable number of plants in each family should be tested, with 20 probably being a minimum. Each family can be tested with several races and this often provides additional information and clarifies the results.

To illustrate the use of backcrosses, data from the backcross of Kenya C9906 to Marquis are given in Table 5.2 (from Knott 1957). With race 56, approximately half the families segregated, indicating that one gene controls resistance. The segregation within families was clear – 480 plants with IT 0;1 and 179 with IT 4, a reasonable fit to a 3:1 ratio, confirming that resistance was controlled by one gene that conditioned an IT 0;1. All 32 families that segregated with race 56 also segregated with race 15B-1, indicating that the same gene gave resistance to both races. However, the families could be divided into two groups, 17 that segregated in a ratio of 1(IT 0;1) : 3(IT 4), and 15 that segregated 4(IT 0;1) : 9(IT $1^{+}3^{y}$) : 3(IT 4). The intermediate class was clearly distinguishable because of a typical yellowing around the pustules and often at the leaf tips and margins as well. In addition, of the 44 families that were

Table 5.2. The results of seedling stem rust tests with races 15B-1 and 56 on F_2 families from the backcross of Kenya C9906 to Marquis. (Knott 1957)

Race 56	Race 15B-1 Number of families 1VR:3S	4VR:9MR:3S	3MR:1S	S	Totals (race 56)	Expected (1:1)	P
3VR:1S	17	15	0	0	32[a]	38	0.10–0.25
S	0	0	21	23	44	38	
Totals (race 15B-1)	17[b]	15[c]	21[d]	23	76	76	
Expected (1:1:1:1)	19	19	19	19	76	76	
P		0.50–0.70					

[a] Race 56 – the segregation within families was 480(IT 0;1): 179(IT 4), P for a 3:1 ratio is 0.10–0.25.
[b] Race 15B-1 – the segregation within families was 81(IT 0;1): 259(IT 4), P for a 1:3 ratio is 0.50–0.70.
[c] Race 15B-1 – the segregation within families was 90(IT 0;1): 188(IT 1^+3^y): 80(IT 4), P for a 4:9:3 ratio is 0.10–0.25.
[d] Race 15B-1 – the segregation within families was 396(IT 1^+3^y): 152(IT 4), P for a 3:1 ratio is 0.10–0.25.

susceptible to race 56, 21 segregated in a ratio of 3(IT 1^+3^y) : 1(IT 4) and 23 were susceptible. Clearly, a second gene conditioning an IT 1^+3^y only to race 15B-1 was segregating. Overall, with race 15B-1 the observed segregation of 17:15:21:23 was a good fit to the expected 1:1:1:1.

Data from F_2 populations and F_3 families have frequently been used in studying the inheritance of rust resistance. The data are more complex than those from backcrosses. In the F_3 generation there are three basic types of families, homozygous resistant, segregating, and homozygous susceptible, compared to only two in a backcross F_2 generation (segregating and homozygous susceptible). If resistance involves two or more dominant genes, it may be difficult to distinguish some segregating families from homozygous resistant families (i.e., distinguishing a family segregating 15:1 or 63:1 from a homozygous resistant family). Certainly, reasonably large numbers of plants per family will have to be grown. As in backcrosses, the segregating families can be of different types depending on the number of genes involved. Despite the somewhat more complex nature of F_3 data, reliable results can often be obtained (see, for example, Berg et al. 1963).

All of the discussion so far has been based on the assumption that the genes under study are not linked. Fortunately, this is often the case, but occasionally linkage will occur and change the ratios. Linkage is most easily detected and measured in backcrosses.

In their original publication on the identification of stem rust races, Stakman and Levine (1922) classified ITs 0, 1 and 2 as resistant and 3 and 4 as susceptible. This division has sometimes been used blindly in genetic studies. For example, plants with an IT of 2^+ or lower are called resistant and those with an IT of 3^- or higher are called susceptible, when in reality the distribution is probably continuous. Therefore, the classification does not reflect genotypes and is meaningless for genetic analysis. Only when the genotypes are clearly reflected in distinct phenotypes will the genetic analysis be valid.

Classifications based on individual plants are even more difficult in the field where factors such as microenvironment, other disease and insect pests, and race mixtures can affect results. Generally, genetic analysis should be based on whole families, not individual plants. Even then problems can arise. For example, if families differ in maturity their reaction to rust can be affected. An early family may escape rust while a late one is subjected to a particularly heavy spore load.

5.4 The Inheritance of Resistance to Stem Rust

Of the five genes for stem rust resistance named by Ausemus et al. (1946), *Sr1* was later found to be an *Sr9* allele and designated *Sr9d* (Knott 1971a) (Table 5.3). The gene *Sr2*, derived from Yaroslav emmer and present in Hope, H44-24 and many of their derivatives, conditions adult plant resistance to many races and also acts as a modifier of seedling resistance. It has been used widely in wheat breeding. Both *Sr3* and *Sr4* were originally described in Marquillo but neither has been identified further or isolated in single gene lines. The gene *Sr5* is present in the standard differential, Reliance, and in Kanred winter wheat, from which it was transferred to Thatcher and derivatives of Thatcher. It conditions immunity or near immunity but to only a limited number of races.

Knott and Anderson (1956) named genes from *Sr6* to *Sr12*. The gene *Sr6* conditions a fleck infection type to many races. It is temperature-sensitive and resistance breaks down between 20 and 22°C, the infection type rising from 0; to X to 4. To some races *Sr6* is dominant and to others recessive but this depends on the genetic background it is in. It has been used fairly extensively in breeding, particularly where temperatures are not too high, at least at night. Virulent races are not uncommon.

The gene *Sr7a* is frequently present in Kenyan cultivars and lines derived from them. The typical infection type is quite variable, ranging from 1 to 3c, usually with considerable yellowing around the pustule and along the leaf margins. Lines

Table 5.3. The identified genes for stem rust resistance, compiled mainly from Roelfs and McVey (1979), McIntosh (1983), Roelfs (1985), and the author's own data

Gene	Common sources	Typical seedling IT	Chromosome location
Sr1 = *Sr9d*			
Sr2	Hope, H44-24, Hopps	4[a]	3BS
Sr3, *Sr4*	Marquillo – not available in separate lines	–	–
Sr5	Kanred, Reliance, Thatcher, Chris, Manitou, Hochzucht	00;	6D
Sr6	Kenyan lines (e.g., Kenya 58), Red Egyptian, Africa 43, Eureka, McMurachy, Kentana 52, Chris, Manitou, Selkirk, Gamut	0;	2Dα
Sr7a	Egypt Na101, many Kenyan lines, Kentana 52	13[c]	4BL
Sr7b	Marquis, Hope, Spica, Renown, Selkirk, Chris, Manitou, Khapstein	2	4BL
Sr8a	Red Egyptian, Mentana, Frontana, Magnif G, Rio Negro	2	6Aα

Table 5.3. *(Cont'd.)*

Gene	Common sources	Typical seedling IT	Chromosome location
Sr8b	Barleta Benvenuto, Klein Titan, Klein Cometa	X	6Aα
Sr9a	Red Egyptian	$1^{+}2^{-}$	2BL
Sr9b	Many Kenyan lines (e.g., Kenya 117a, Kenya Farmer), Frontana, Gamenya, Festival, Gamut	2^{-}	2BL
Sr9c	Reserved for *SrTtl* (*Sr36*) which was later found not to be an *Sr9* allele		
Sr9d	Hope, H-44-24, Lancer, Scout, Lawrence, Renown, Redman	$0;2^{-}$	2BL
Sr9e	Vernstein, Vernal emmer	0;1	2BL
Sr9f	Chinese Spring	0;2	2BL
Sr9g	Thatcher, Kubanka, Acme	22^{+}	2BL
Sr10	Egypt Na95, Kenyan lines	0;X	–
Sr11	Lee, Gabo, Kenya Farmer, Charter, Sonora 64, Tobari 66, Yalta, Mendos	12^{-}	6BL
Sr12	Thatcher, Windebri, Egret, Chris, Manitou	X	3BS
Sr13	Khapstein, Madden	$2^{-}2$	6Aβ
Sr14	Khapstein	12^{n}	1BL
Sr15	ASII, Axminster, Festival, Norka, Thew, Normandie	$0;1^{+n}$	7AL
Sr16	Thatcher, Reliance	2	7AL
Sr17	Hope, Renown, Redman, Lawrence, Spica, Warigo, Aotea	0;X	7BL
Sr18	Marquis, Reliance, many wheat lines	$0;2^{=}$	1D
Sr19	Marquis	1^{-}	2B
Sr20	Marquis, Reliance	$2^{=}$	2B
Sr21	Einkorn (*Triticum monococcum*), tetraploid and hexaploid derivatives	$1^{-}2^{-}$	2A
Sr22	Einkorn (*Triticum monococcum*), tetraploid and hexaploid derivatives	12^{-}	7AL
Sr23	Exchange, Etoile de Choisy, Selkirk, Warden	23^{c}	4A
Sr24	*Agropyron elongatum*, Agent, Blueboy II, Cloud, Fox, Sage, Sears' 3D/Ag translocations	12^{-}	3DL
Sr25	*Agropyron elongatum*, Agatha, Sears' 7D/Ag translocations	12^{-}	7DL
Sr26	*Agropyron elongatum*, Knott's 6A/Ag translocations, Eagle, Kite	12^{-}	6Aβ
Sr27	*Secale cereale*, Wheat-rye translocation WRT 238-5	0;	3A
Sr28	Line AD, Kota	0;	2BL
Sr29	Etoile de Choisy	23^{-}	6Dβ
Sr30	Webster, Festiguay	2^{-}	5DL
Sr31	*Secale cereale*, 1B/1R translocations, Aurora, Kavkaz, Lovrin, Neuzucht, Veery, Weique	0;1	1BL/1RS
Sr32	*Triticum speltoides* derivatives	2^{-}	2A,2B,2D
Sr33	RL 5405 (Tetra Canthatch/*Aegilops squarrosa*)	$1^{+}2$	1DL
Sr34	*Triticum comosum*, Compair, various translocation lines	$1^{+}2$	2A,2D
Sr35	Triticum monococcum derivatives, Arthur	0;	3AL
Sr36	*Triticum timopheevii*, C.I. 12632, C.I. 12633, Cook, Idaed 59, Timgalen, Timvera, formerly *SrTtl*	$0;1^{+}$	2BS
Sr37	Steinwedel/*Triticum timopheevii* derivative	0;	4Aβ

[a] Adult plant resistance only.

carrying *Sr7a* are susceptible to many old races but the gene became important in North America because it conditions resistance to race 15B-1. The *Sr7b* allele is present in both Marquis and Kota of the Stakman differentials. It was discovered when the production of a near-isogenic line of Marquis carrying *Sr7a* resulted in the loss of the Marquis resistance to race 48A. Virulence on *Sr7b* is common and the gene has not been used in breeding.

The gene *Sr8a* is present in many Italian cultivars, from which it spread to South American, Kenyan, and CIMMYT cultivars. Typically it conditions an infection type 2. It provided useful resistance to some of the older North American races, including 15B-1 and 56, but is ineffective against many newer races. In some countries *Sr8a* conditions resistance in seedlings but not in adult plants, even with the same race (Luig 1983). The allele *Sr8b* was recently identified by Singh and McIntosh (1986) in Barleta Benvenuto and Klein Titan. The gene is recessive and ineffective when hemizygous.

The most extensive set of alleles for stem rust resistance occurs at the *Sr9* locus with six alleles already identified. When the locus was first discovered, it was not realized that Red Egyptian carried a different allele than many Kenyan cultivars. However, when two near-isogenic lines of Marquis were produced and tested with a number of North American races, it was found that the gene from Kenya 117A (*Sr9b*) gave resistance to more races than the gene from Red Egyptian (*Sr9a*) (Green et al. 1960). The difference is most striking in Australia, where all isolates are virulent on *Sr9a* but many are avirulent on *Sr9b* (Luig 1983). The symbol *Sr9c* had been reserved for a gene from *T.timopheevii (SrTt1)*, which was later found not to be an *Sr9* allele. The *Sr9d* allele was identified in Hope and H44-24 (Knott 1968) and is also present in three standard differentials, Arnautka, Mindum, and Spelmar. It is present in many Hope and H44-24 derivatives but virulence is common. The *Sr9e* allele was transferred from one of the standard differentials, the tetraploid Vernal emmer, to the hexaploid Vernstein (Luig and Watson 1967). It has not been used to any extent in breeding but virulence is widespread. The *Sr9f* allele is of interest mainly because it is present in Chinese Spring. It is effective against very few stem rust isolates. The final allele, *Sr9g*, is present in two of the standard differentials, the tetraploids Kubanka and Acme. It is also present in Thatcher, which presumably obtained it from Iumillo durum. Very few isolates are avirulent on *Sr9g*.

The gene *Sr10* was identified in a number of Kenyan lines. It conditions a rather variable infection type, often an X. Although it has not been deliberately used in breeding, the widespread use of Kenyan material as parents has resulted in it being present in many cultivars, particularly from the CIMMYT program.

The gene *Sr11* was identified in Lee, Gabo, and Timstein and was thought to be one of a pair of complementary genes. However, Loegering and Sears (1963) found that there was only a single gene for resistance but it was linked to a pollen-killer locus which resulted in distorted segregations in crosses. Although it was originally a valuable gene in many wheat-growing areas, newer races are largely virulent.

Originally *Sr12* was used to designate the gene that was complementary in action to *Sr11*. Later, Sheen and Snyder (1964) used the symbol for a gene in a Marquis selection. The gene is found in Thatcher and some of its derivatives. It provides resistance to race 56, which was prevalent in North America when Thatcher was released.

The two genes *Sr13* and *Sr14* were identified in Khapstein (Knott 1962b) and were derived from the standard differential, Khapli emmer. The gene *Sr13* conditions somewhat variable infection types with the infection type being lower at higher temperature (30°C) (Roelfs and McVey 1979). *Sr14* conditions a 1 to 2 infection type, often with considerable necrosis. Together, *Sr13* and *Sr14* produce a lower infection type. Neither appears to have been used to any extent in breeding.

The gene *Sr15* was discovered by Watson and Luig (1966) and is present in several older cultivars such as Norka and Thew. It is temperature sensitive. In Australia in tests at controlled temperatures, isogenic lines carrying *Sr15* gave five different infection types (Luig 1983). The gene is closely linked to *Lr20* and *Pm1* (powdery mildew resistance).

The gene *Sr16* was found in Thatcher (Loegering and Sears 1966) and gives an infection type 2 to a limited number of races.

McIntosh et al. (1967) identified *Sr17* in Hope and some of its derivatives. The gene is recessive and shows some temperature sensitivity (Roelfs and McVey 1979). It is present in many Hope and H-44-24 derivatives including many cultivars in Australia, the U.S.A., and the CIMMYT program. Virulence is common.

Anderson et al. (1971) identified *Sr18* and *Sr20* in both Marquis and Reliance, and *Sr19* in Marquis. Avirulence on each of the three genes is rare (Roelfs and McVey 1979). *Sr18* is present in many cultivars, including derivatives of Hope and H-44-24, which undoubtedly obtained it from Marquis.

Both *Sr21* and *Sr22* were transferred first from diploid einkorn wheat to tetraploid wheat by Gerechter-Amitai et al. (1971) and then by The (1973) to hexaploid wheat. Kerber and Dyck (1973) also transferred *Sr22* from a diploid to a tetraploid to a hexaploid wheat. *Sr21* is present in the standard differential, Einkorn, but conditions resistance to relatively few races. However, it does give resistance to race 56, which has been important in North America for many years. *Sr22* is effective against most races but lines carrying it may be moderately susceptible in the field (Roelfs and McVey 1979). Neither *Sr21* nor *Sr22* appears to have been used in commercial cultivars.

The gene *Sr23* is present in Warden, from which it was transferred to Exchange and then to Selkirk (McIntosh and Luig 1973). It gives only moderate resistance to a limited number of races, mostly in North America. It is closely linked to *Lr16*.

The genes *Sr24, Sr25* and *Sr26* were all transferred to wheat from tall wheatgrass (*Agropyron elongatum = Elytrigia pontica*). *Sr24* involves a natural translocation between an *Agropyron* chromosome and wheat chromosome 3D in which whole arms may have been exchanged (Smith et al. 1968; Gough and Merkle 1971; McIntosh et al. 1977). This line was named Agent. The gene *Lr24* is also on the translocated segment of *Agropyron* chromosome. The two genes are present in a number of cultivars in the U.S.A. *Sr24* is apparently effective against all races of stem rust but the resistance due to *Lr24* has broken down. Sharma and Knott (1966) transferred leaf rust resistance (*Lr19*) from a wheat-*Agropyron* derivative to wheat, and transferred *Sr25* and a gene for yellow pigment on the same segment of an *Agropyron* chromosome. Seedlings carrying *Sr25* are resistant to almost all races of stem rust but the resistance is only moderately effective in adult plants in the field (McIntosh et al. 1977; Roelfs and McVey 1979). The resistance has not been used in commercial cultivars because of the undesirable yellow flour pigment, but Knott

(1980) produced two mutant lines lacking the gene. The third *Agropyron* gene, *Sr26*, was translocated to wheat by Knott (1961). It has been used in a number of Australian cultivars and appears to be resistant to all races of stem rust in the field.

The gene *Sr27* was transferred to wheat from rye by Acosta (1963). It conditions a fleck reaction to all races of wheat stem rust but is susceptible to some races of rye stem rust. The gene has not been used in commercial cultivars of wheat but it is widespread in triticales (McIntosh et al. 1983).

McIntosh (1978) identified *Sr28* in Kota. The gene is effective against very few races.

The gene *Sr29* was discovered in Etoile de Choisy by McIntosh et al. (1974). Seedlings carrying *Sr29* are moderately resistant to all races but adult plants are moderately susceptible (Roelfs and McVey 1979).

Stakman et al. (1925) found that Webster, an introduction to the U.S.A. from Russia, had comprehensive resistance to stem rust. Knott and McIntosh (1978) showed that its resistance was due to a nearly recessive gene, *Sr30*. It gives moderate resistance to many races but seems to be present in only a few Webster derivatives.

The gene *Sr31* is derived from rye and is present in wheat as either a substitution of rye chromosome 1R for wheat chromosome 1B or as a 1BL/1RS translocation that occurred naturally (Mettin et al. 1973; Zeller 1973). The segment of rye chromosome 1R also carries *Lr26* and *Yr9* and has been used widely in European cultivars. *Sr31* provides good resistance to many races.

The gene *Sr32* was identified by Sears (unpublished) in a *T. speltoides* derivative (McIntosh 1983).

Kerber and Dyck (1979) produced a synthetic hexaploid wheat by crossing Tetra Canthatch (AABB) with *Aegilops squarrosa* (DD). They showed that it carried a single gene, later designated *Sr33*.

The gene *Sr34* was transferred to wheat from *Aegilops comosa*, along with *Yr8* (McIntosh et al. 1982).

The gene *Sr35* was transferred from diploid einkorn wheat to tetraploid wheat and then hexaploid wheat independently in Australia and Canada (McIntosh et al. 1984). It gives a low infection type but many races are virulent on it.

Both *Sr36* and *Sr37* were transferred to wheat from *T. timopheevi*. The gene that was first designated *SrTt1* and then *Sr36* was transferred by Allard and Shands (1954) along with gene *Pm6* for powdery mildew resistance. It has been used widely in breeding and is present in a number of Australian and American cultivars. With some races, *Sr36* conditions as unusual infection type, a combination of type 4 and 0; pustules (Ashagari and Rowell 1980). Virulence on *Sr36* is common. *Sr37*, which was first designated *SrTt2*, was transferred to bread wheat by Gyarfas (McIntosh and Gyarfas 1971). It provides resistance to most races of stem rust but appears not to have been used to any extent in breeding.

In addition to the designated *Sr* genes, a number of other genes have been identified but not yet given *Sr* numbers because their relationship with the named genes is not known. Roelfs (1985) and Luig (1983) list a number of them.

5.5 The Inheritance of Resistance to Leaf Rust

Browder (1980) reviewed much of the information about genes for resistance to leaf rust.

Ausemus et al. (1946) designated three genes for leaf rust resistance, *Lr1, Lr2* and *Lr3,* based on the work of Mains et al. (1926). *Lr1* was identified in Malakof and is present in a number of cultivars (Table 5.4). Virulence on *Lr1* is common in North America (Browder, 1980). A gene in Webster was designated *Lr2* and became *Lr2a* when other alleles were identified at the locus. In a study of leaf rust resistance in the eight standard differential cultivars, Soliman et al. (1964) found that Webster, Carina, Brevit, and Loros carried alleles at the *Lr2* locus and designated them as *Lr2,* $Lr2^2$, $Lr2^3$, and $Lr2^4$. Later the symbols were changed to *Lr2a, Lr2b, Lr2c,* and *Lr2d.* Finally, Dyck and Samborski (1974) showed that *Lr2c* and *Lr2d* were the same and the latter symbol was dropped. Ausemus et al. (1946) assigned the symbol *Lr3* to a gene in Fulcaster that was not detectable in seedlings. However, Soliman et al. (1964) used *Lr3* to designate a gene for seedling resistance in Mediterranean and Democrat, which is not the gene in Fulcaster (Browder 1980). Their designation is the accepted one. Haggag and Dyck (1973) identified two *Lr3* alleles, *Lr3ka* in Klein Aniversario and *Lr3bg* in Bage. The original allele became *Lr3a.* Haggag and Dyck's (1973) data indicated that Prelude carries a suppressor that makes *Lr3a* ineffective against some leaf rust races.

Fitzgerald et al. (1957) used the symbols *Lr4* to *Lr8* for five genes identified in a highly resistant, soft red winter wheat selection, Purdue 3369-61-1-1-10-8 (Waban). However, their relationship to other genes is not known and near-isogenic lines are not available (Browder 1980).

The gene *Lr9* was transferred to wheat from *Aegilops umbellulata* by Sears (1956). It has been used in a number of cultivars in the U.S.A. but virulence on it is now moderately frequent (Browder 1980).

Anderson (1961) identified a gene, *L,* in several cultivars including Lee, Gabo, Mayo 52, Exchange and Selkirk. Dyck and Samborski (1968b) designated it as *Lr10.* The gene has been widely used in commercial cultivars but virulence on it is now common in North America (Browder 1980).

The gene *Lr11* was identified by Soliman et al. (1964) in Hussar. It is less effective at high temperature (30°C) (Williams and Johnston 1965).

Dyck et al. (1966) identified two genes conditioning adult plant resistance, *Lr12* in Exchange and *Lr13* in Frontana. *Lr12* is partially dominant while *Lr13* varies from recessive to partially dominant in different crosses. Both are subject to modifying genes. The resistance due to *Lr13* becomes effective at about the third leaf stage while *Lr12* is most effective at the flag leaf stage. Both genes were used in commercial cultivars in North America but virulence then became common. However, Rajaram et al. (1988) emphasize the importance of *Lr13* in combination with other genes in producing durable resistance to leaf rust in CIMMYT cultivars.

McIntosh et al. (1967) identified an incompletely dominant gene for leaf rust resistance in Spica, which was linked to genes for stem rust resistance (*Sr17*-18% recombination) and powdery mildew resistance. The gene, *Lr14,* is present in many Hope and H-44-24 derivatives. Dyck and Samborski (1970) transferred genes for leaf rust resistance from Selkirk (*Lr14*), Maria Escobar, and Bowie into Thatcher.

Table 5.4. The identified genes for leaf rust resistance, compiled mainly from Browder (1980), McIntosh (1983), and Singh (1984)

Gene	Common sources	Typical seedling IT	Chromosome location
Lr1	Malakof, Blueboy, Centenario, Sonora	0;	5D
Lr2a	Webster, Eureka, Waldron, Festiguay	0;1	2D
Lr2b	Carina	0;1	2D
Lr2c	Brevit, Loros	0;1	2D
Lr3a	Mediterranean, Democrat, Bowie, Pawnee, Warrior	0;1	6BL
Lr3ka	Klein Aniversario, Klein Titan	1+2	6BL
Lr3bg	Bage	0;1	6BL
Lr4-Lr8	Purdue 3369-61-1-1-10 (Waban)	–	–
Lr9	*Aegilops umbellulata*, Transfer, Abe, Arthur 71, McNair 701, Riley 67, Oasis	0;	6BL
Lr10	Lee, Exchange, Gabo, Selkirk, Mayo 54, Blueboy	0;1	1AS
Lr11	Hussar, Bulgaria 88, Oasis	12	2A
Lr12	Exchange, Opal	4[a]	4A
Lr13	Frontana, Chris, Manitou, Neepawa, Era, Polk, Egret	0;-X	2BS
Lr14a	Spica, Hope, Selkirk, Aotea, Glenwari, Hofed	X	7BL
Lr14b	Maria Escobar, Bowie, Rafaela	X	7BL
Lr15	Kenya 1-12-E-19-J	0;	2Dα
Lr16	Exchange, Etoile de Choisy, Warden, Selkirk, Columbus	12	4A
Lr17	Klein Lucero, Maria Escobar, Jupateco, Rafaela	0;1	2AS
Lr18	Africa 43, Red Egyptian P.I. 170925, Timvera	12	5BL
Lr19	*Agropyron elongatum*, Agatha	0;	7DL
Lr20	Thew, Axminster, Festival, Kenya W744, Normandie	0;	7AL
Lr21	Tetra Canthatch/*Aegilops squarrosa* var. *meyeri*	0;12	1DL
Lr22a	Tetra Canthatch/*Aegilops squarrosa* var. *strangulata*	4[a]	2D
Lr22b	Thatcher	4[a]	2D
Lr23	Gabo, Lee, Kenya Farmer, Gamenya, Timstein	0;1	2BS
Lr24	*Agropyron elongatum*, Agent, Blueboy II, Cloud, Fox, Sage, Sears, 3D/Ag translocations	0;	3DL
Lr25	*Secale cereale*, Transec, Transfed	0;12	4Aß
Lr26	*Secale cereale*, 1B/1R translocations, Aurora, Kavkaz, Lovrin, Neuzucht, Veery, Weique	0;	1BL-1RS
Lr27	Gatcher, Hope, Ciano, SUN 27A, complementary to *Lr31*	X=	3BS
Lr28	*Triticum comosum*, CS 2A/2M 4/2, CS 2D/2M 3/8	0;	4BL
Lr29	*Agropyron elongatum*, Sears 7D/Ag#11	0;1	7DS
Lr30	Terenzio	1+2	4BL
Lr31	Chinese Spring, Hope, Ciano, SUN 27A, Gatcher, complementary to *Lr27*	X=	4Aß
Lr32	Tetra Canthatch/*T. tauschii* RL5497-1	–	3D
Lr33	RL6057 (=Tc*6/P.I. 158458)	1+	1BL
Lr34	Lageadinho, Terenzio, complementary to *Lr33*	22+	7D

[a]Adult plant resistance.

Crosses indicated that the genes in Maria Escobar and Bowie were the same and were closely linked to *Lr14*. They obtained one recombinant in 644 testcross plants but concluded that the two genes were alleles and designated them *Lr14a* (Selkirk and Spica) and *Lr14b*. In the pathogen, virulence on *Lr14a* was recessive and on *Lr14b* was dominant. The two genes for virulence were inherited independently.

Luig and McIntosh (1968) identified a gene for leaf rust resistance in Kenya 112-E-19-J (L), located it on chromosome 2D and designated it *Lr15*. The gene was closely linked to *Sr6* (no recombinants in 100 F_3 lines) and more loosely with *C* (compact head, 26% recombination), two genes known to be on chromosome 2D. Luig and McIntosh (1968) also found that a gene for leaf rust resistance in Festiguay, which appeared to be the same as one in Webster (*Lr2a*), was closely linked to *Sr6* and loosely linked to *C* (26% recombination). They reported that a cross between Kenya 112 and Webster segregated, but McIntosh and Baker (1968) obtained no recombinants in the same cross. If the gene they studied in Webster is *Lr2a*, then it is on chromosome 2D and possibly allelic to *Lr15*. This conflicts with Soliman et al.'s (1964) results, which indicated that *Lr2a* is on chromosome 1B. However, the evidence for chromosome 2D is much stronger.

A gene designated *LrE* was reported by Anderson (1961) in Exchange and Selkirk and later designated *Lr16* by Dyck and Samborski (1968b). It is present in the Canadian cultivar Columbus (Samborski and Dyck 1982).

Dyck and Samborski (1968b) produced near-isogenic lines of Thatcher and Prelude, carrying single genes from Klein Lucero, Maria Escobar, P.I. 170925, P.I. 170916-2c, and Africa 43. The lines were grouped according to their reaction to six leaf rust races, intercrossed and crossed to lines carrying *Lr1, Lr2, Lr3, Lr10,* and *Lr16.* Based on the results, they identified a gene, *Lr17,* in Klein Lucero and Maria Escobar, and a second gene, *Lr18,* in P.I. 179025, P.I. 170916-2c, and Africa 43. Both are partially dominant and temperature-sensitive.

Three genes for leaf rust resistance have been transferred to wheat from *Agropyron elongatum* (= *Elytrigia pontica*). In the case of *Lr19,* the primary objective was the transfer of leaf rust resistance from Agrus, a wheat-*Agropyron* derivative (Sharma and Knott 1966), and the presence of stem rust resistance (*Sr25*) was discovered later (see Sect. 5.4). Although *Lr19* appears to condition resistance to all races of leaf rust with which it has been tested, it has not yet been used in cultivars because of a linked gene for yellow pigment in the flour. However, Weibull in Sweden have released a cultivar (Sunnan) carrying the gene for yellow pigment. It is intended for use in producing pasta products in which yellow color is desirable. A second *Agropyron* gene, *Lr24,* was transferred to the cultivar Agent, along with *Sr24* (Sect. 5.4). It has been used in a number of cultivars but resistance has broken down. Sears (1973) produced several transfers of leaf rust resistance from the *Agropyron* chromosome in Agrus to wheat chromosome 7D, using induced homoeologous pairing. One of these lines appeared to carry a gene for leaf rust resistance derived from the opposite arm of the *Agropyron* chromosome than *Lr19* was. It conditions a higher infection type ($0;^{n}$ versus 0;) and is not linked to the gene for yellow pigment. It has been designated *Lr29.*

Watson and Baker (1943) reported that Thew carries a dominant gene for leaf rust resistance which was linked to a gene for resistance to powdery mildew (later identified to be *Pm1*). Watson and Luig (1966) showed that the gene was also linked to a gene for stem rust resistance. Browder (1972) designated the gene as *Lr20.*

Kerber and Dyck (1969) and Dyck and Kerber (1970) produced two synthetic hexaploids by combining Tetra Canthatch (2n = 28, AABB) with two varieties of *Aegilops squarrosa,* var. *meyeri* and var. *strangulata.* The first proved to carry a partially dominant gene linked to brown glumes. The second synthetic hexaploid carried a partially dominant gene for adult plant resistance that was independent of *Lr12* and *Lr13*, the previously identified genes for adult plant resistance. It was linked to genes for threshability (6.0% recombination) and waxiness (15.6% recombination). Rowland and Kerber (1974) located the first gene on chromosome 1D and designated it *Lr21*, and the second gene on 2D and designated it *Lr22*. When *Lr22* was transferred to Thatcher, the backcross lacked the Thatcher type of leaf rust resistance. Dyck (1979) showed that the adult plant resistance of Thatcher was due to an allele of *Lr22* and named it *Lr22b* and the original allele *Lr22a*.

A gene for leaf rust resistance derived from Gaza durum (Watson and Luig 1961) was given the temporary designation *LrG*. When it was released in the Australian bread wheat, Gabo, virulence appeared 4 years later. McIntosh and Dyck (1975) showed that the gene was on chromosome 2B linked to *Sr9* (22% recombination), and renamed it *Lr23*. It varied from recessive to partially dominant in different crosses and the dominance was greater at higher temperatures.

Two genes for leaf rust resistance have been transferred to wheat from rye. Driscoll and Jensen (1963, 1964) used irradiation to translocate a piece of a rye chromosome carrying genes for leaf rust and powdery mildew resistance to wheat chromosome 4A. The gene for leaf rust resistance (*Lr25*) is partially dominant and is expressed more strongly in adult plants than in seedlings. Although it is effective against most, if not all, leaf rust races, it has not been used in commercial cultivars, presumably because of deleterious effects from the rye chromatin. The second gene from rye is *Lr26*, which is present in many cultivars which have either a substitution of rye chromosome 1R for wheat chromosome 1B or a natural translocation, 1BL/1RS (see Sect. 5.4). In this case the rye chromosome or segment did not have deleterious effects on agronomic characters but did affect dough characteristics, resulting in a sticky dough (Dhaliwal et al. 1987).

Complementary genes for leaf rust resistance were identified by Singh and McIntosh (1984a) in Gatcher and certain other wheats. Chinese Spring carries one of the genes and Hope and many of its derivatives carry the second which is closely linked to *Sr2*. Singh and McIntosh (1984b) showed that the gene in Chinese Spring is on chromosome arm 4AB and labeled it *Lr31*. The gene in Hope and its derivatives was located on chromosome arm 3BS. It had previously been identified in Gatcher and designated as *Lr27*.

When Riley et al. (1968a) transferred yellow rust resistance (*Yr8*) from *Aegilops comosa* to wheat, they used a cross with *Aegilops speltoides* to induce homoeologous pairing between wheat and *comosa* chromosomes. Homoeologous pairing also occurred between wheat and *speltoides* chromosomes and McIntosh et al. (1982) showed that two of the lines with yellow rust resistance also carried leaf rust resistance derived from *Ae. speltoides*. The gene was designated *Lr28*.

Samborski and Dyck (1976) transferred a recessive gene for leaf rust resistance from Terenzio to Thatcher and designated it *LrT*. Dyck and Kerber (1981) located it on chromosome 4B and renamed it *Lr30*.

The gene *Lr32* was reported by Kerber (unpublished) in a Tetra Canthatch/*Triticum tauschii* (= Ae. squarrosa) synthetic hexaploid (McIntosh et al. 1987).

Dyck (1977) detected two genes for leaf rust resistance in PI 58548, a bread wheat introduction from China, and produced near-isogenic lines of Thatcher carrying them. The two genes interacted to give a lower infection type than either alone. Dyck et al. (1987) located one of the genes on the long arm of chromosome 1B and designated it *Lr33*. It is closely linked to *Lr26* (2.6 ±0.8% recombination) or to the translocation breakpoint, which is probably the centromere.

Dyck and Samborski (1982) identified two genes (*LrT2* and *LrT3*) in a number of bread wheats, including Terenzio and Lageadinho. The two genes conditioned moderate resistance by themselves but interacted to give good resistance (0;X). The first gene has been designated *Lr34* (McIntosh et al. 1987). It interacts with *Lr33* to give increased resistance and is most readily detected in adult plants in the field. When *Lr34* was transferred to Thatcher, the backcross line proved to be more resistant to stem rust than Thatcher. Dyck and Samborski (1982) suggested that the *Lr34* locus in Thatcher was linked to an inhibitor which was lost in the production of the backcross line. The absence of the inhibitor allowed the expression of genes for stem rust resistance in Thatcher. In further testing, Dyck (1987) showed that *Lr34* was on chromosome 7D and was either a noninhibiting allele of the inhibitor or was closely linked to a noninhibitor.

5.6 The Inheritance of Resistance to Yellow Rust (with R. Johnson)

As already noted, the first evidence that rust resistance could be controlled by a single Mendelian gene was presented by Biffen (1905) working with yellow rust. In crosses involving Rivet wheat, resistance was recessive if F_2 plants were scored late in development and all plants with more than a trace of rust were included in the susceptible class. In fact, one quarter of the plants rusted rapidly, while half of them rusted more slowly, indicating incomplete dominance. Rivet is a tetraploid wheat, which was probably not known to Biffen. In crosses with a susceptible hexaploid cultivar, variable chromosome numbers and considerable sterility would be expected in the F_2 and subsequent generations in addition to segregation for resistance to yellow rust (see Robbelen and Sharp 1978). This may have contributed to the considerable losses of plants from an original 300 F_1 grains down to only 163 lines in the F_3, many of them with only a few plants. The number of susceptible lines was below the expected value for a single recessive gene, although this was attributed by Biffen to poor reproduction of the highly susceptible lines. Perhaps more important than the precise genetical description of the control of resistance was the demonstration by Biffen that resistance was inherited independently of some other characters and that lines breeding true for resistance could readily be obtained, thus indicating the possibilities of breeding for resistance. In those early years he was unaware of the potential variation in virulence of *P. striiformis* and believed that rusts would be reliably controlled by breeding resistant varieties (Biffen 1911-12, 1931).

Following the work of Biffen, many studies were conducted on the resistance of wheat to yellow rust, and much of the information is reviewed by Robbelen and Sharp (1978). As with the other rusts, resistance in seedlings was often attributed to one or a few genes. Unfortunately, there was little co-ordination between workers and no formal system of nomenclature of resistance genes until Lupton and Macer

(1962) proposed the use of the symbol *Yr*. Much of the earlier work has never been incorporated into this system and interrelationships between many resistance genes remain unknown. The work of Manners (1950) and Zadoks (1961) demonstrated that resistance developed by plants after the seedling stage could be race-specific. Zadoks (1961) tested the reactions of 17 wheat cultivars to 15 biotypes of yellow rust and, based on the application of the gene-for-gene system as proposed by Person (1959), classified 11 of the cultivars into six groups. The validity of the groups was not tested by crossing.

Lupton and Macer (1962) designated four independent loci *Yr1* to *Yr4* controlling resistance in seedlings, including three proposed alleles at the *Yr3* locus and two at the *Yr4* locus (Table 5.5). Macer (1966) added a further three loci, *Yr5* to *Yr7*. The gene *Yr8* was introduced to wheat from *Aegilops comosa (Triticum comosum)* by Riley et al. (1968a). Macer (1975) designated a gene present in lines carrying the wheat-rye translocated chromosome 1BL/1RS as *Yr9* and a gene in the cultivar Moro as *Yr10*. Genes *Yr11* to *Yr14* were designated by McIntosh (1986) and control a set of race-specific interactions observed in adult plants of several cultivars of wheat with isolates of yellow rust collected in the United Kingdom. The gene *Yr15* was derived from *Triticum dicoccoides* and is effective in seedlings (Gerechter-Amitai et al., unpublished manuscript). The gene *Yr16* controls adult plant resistance in Cappelle-Desprez and was identified by use of chromosome substitution and the development of homozygous recombinant lines (Worland and Law 1986). The

Table 5.5. The identified genes for yellow rust resistance, compiled mainly from McIntosh (1983, 1986) and supported by data from the Plant Breeding Institute, Cambridge

Gene	Common source	Typical seedling IT	Chromosome location
Yr1	Chinese 166, Heines 110, Tadorna, Maris Templar, Hustler	00	2A
Yr2	Heines VII, Merlin, Soissonais-Desprez, Tadorna, Hustler	0–2	7B
Yr3a	Cappelle-Desprez, Hobbit, Maris Ranger, Maris Templar	0–1+	–
Yr3b	Hybrid 46	0–1+	–
Yr3c	Minister	0–1+	–
Yr4a	Cappelle-Desprez, Hobbit, Maris Ranger, Maris Templar	0–1+	–
Yr4b	Hybrid 46	0–1+	–
Yr5	*Triticum spelta album*	00–0	2BL
Yr6	Heines Kolben, Peko	0–2+	7BS
Yr7	Thatcher, Lee, Nudif TP241, Reichersberg 42, Flevina	ON-1+N	2BL
Yr8	Compair (from *T. comosum*)	00	2D
Yr9	1BL/1RS wheat-rye translocations, Neuzucht, Weique, Kavkas, Veery	00–0	1BL/1RS
Yr10	Moro, PI 178383	00	1BS
Yr11	Jose Cambier	A	–
Yr12	Mega, Armada	A	–
Yr13	Maris Huntsman, Maris Nimrod, Mardler, Sportsman, Hustler	A	–
Yr14	Maris Bilbo, Hobbit, Avalon	A	–
Yr15	Hexaploid derivatives of *T. dicoccoides* G-25	00	1B*
Yr16	Cappelle-Desprez	A	2DS

A = adult plant resistance.
*R. A. McIntosh personal communication.

genes *Yr1* to *Yr14* are all known to be race-specific but as yet there is no evidence for race specificity of *Yr15* and *Yr16*.

Many other examples of race-specific resistance to yellow rust have been observed and many of them are likely to be controlled by genes other than those so far designated. For example, Stubbs (1985, Table IV) numbered race-specific host differential cultivars from 0 to 15. For the numbers 1 to 10 these corresponded approximately with the named genes, but for the numbers 11 to 15 the specificities were different from those designated by the genes *Yr11* to Yr15.

There are several unresolved problems relating to the genes named so far. Lupton and Macer (1962) proposed that there was allelism or close linkage of genes in Cappelle-Desprez (CD), Hybrid 46 (H46), and Minister (M). For resistance to race 5 of yellow rust they proposed a single gene in each of M and H46 and two genes in CD. The cross between CD and H46 did not segregate susceptible plants. Susceptible plants were obtained from the cross M × H46 and, contrary to their statement, also in the cross CD × M. They proposed that the single gene in H46 was allelic with one of the genes in CD, and called the allele in CD *Yr4a* and in H46 *Yr4b*. Despite the segregation of susceptible plants, the gene in M was proposed to be allelic with the other gene in CD; that in CD was called *Yr3a* and that in M was called *Yr3c*. Resistance to race 8 was stated to be due to one gene in both CD and M, and to two genes in H46. There was no segregation in the crosses CD × M and CD × H46, and only two susceptible plants out of 147 in the F_2 of H46 × M. It was proposed that the single gene in CD was *Yr3a*, the gene in M was *Yr3c*, and that in H46 there was a gene closely linked or allelic to the *Yr3* locus, and called *Yr3b*. Their proposal for an allelic gene in M at the *Yr3* locus requires further investigation as their data indicated segregation of some susceptible plants from crosses with H46 with each of the three races used, and with one of the three races used to test crosses with CD.

Many cultivars related to CD or to H46 show similar specificity to these two cultivars in their reactions to current United Kingdom races of yellow rust. Using the technique of producing inbred backcross lines (Wehrhan and Allard 1965), Johnson (1980) suggested that Maris Beacon, which gives similar reactions to H46, possessed a single gene. Worland and Law (1987) used chromosome substitution and reciprocal monosomic analysis to identify a single dominant gene on chromosome 3B in the cultivar Avalon, which also reacts similarly to H46. In both these tests race 41 E136 of yellow rust with pathogenicity for CD but not for H46 was used. No tests for allelism of this gene with the genes in CD could be carried out using this race. In tests with race 37 E132 there was evidence of allelism or close linkage of resistance in Avalon with resistance in CD, and that this resistance was not controlled by chromosome 3B (Worland and Law 1987). However, because the races are not the same as those used by Lupton and Macer (1962) it is not at present possible to conclude how these genes relate to those they described for H46 and CD. One feature of the race pattern of yellow rust in Britain may support the suggestion of a gene common to CD and H46 plus an additional gene in H46. Races lacking pathogenicity for both cultivars and others possessing pathogenicity for both are known, as well as races with pathogenicity for CD but not for H46. However, no race with pathogenicity for H46 but lacking pathogenicity for CD has been reported. This would be expected if races with pathogenicity for H46 required pathogenicity for the gene in common with CD as well as for the gene on chromosome 3B.

The genes *Yr5* and *Yr7* are usually dominant and both are located on the long arm of chromosome 2B (Table 5.5.). They were shown to be close to the centromere and possibly allelic (Gaines 1976). Johnson and Dyck (1984) tested F_1 plants from crosses of *T. spelta* (*Yr5*) with Thatcher or Lee, both of which possess *Yr7*, with race 43 E138 which possesses pathogenicity for *Yr7* but not for *Yr5*. The F_1 with Lee was resistant but that with Thatcher was susceptible despite the presence of heterozygous *Yr5*. In the cross of *T. spelta* with Thatcher, segregation of susceptible plants occurred in the F_2 and F_3 generations when tested with a race lacking pathogenicity for both genes. Johnson and Dyck (1984) suggested that Thatcher may possess a dominant inhibitor of the gene *Yr5*. There were no susceptible plants in an F_2 of 200 plants and an F_3 of 200 families from the cross of *T. spelta* (*Yr5*) with Lee (*Yr7*), suggesting that Lee does not possess the inhibitor and supporting the evidence of linkage or allelism between these genes (Johnson et al. 1986). The presence of an inhibitor of *Yr5* in the cultivar Thatcher may explain why, when Macer (1966) first identified *Yr5* and *Yr7*, he did not propose linkage or allelism between them.

The genes *Yr11* to *Yr14* were designated without the usual supporting genetic data because the specificities they relate to were widely publicized as resistance factors R11 and R14 (Priestley 1978). Their designation as genes precludes the assignation of these numbers to other unrelated genes unless there is evidence of allelism. Evidence from races of yellow rust which distinguish these four resistances from combinations of the named race-specific genes in existing cultivars (McIntosh 1986) and data from crosses involving *Yr11*, *Yr13*, and *Yr14* (Wallwork 1982), support the hypothesis that they are single distinct genes that interact on a gene-for-gene basis with specific pathogenicity in *P. striiformis.*

In addition to the many examples of race-specific resistance to yellow rust, there is much other resistance that does not display race specificity with currently available races. Various studies of the genetic basis of some of this resistance have been conducted but individual genes have not been identified, and much of the resistance has not yet been studied genetically.

5.7 Identifying an Unknown Gene for Rust Resistance

As the number of identified loci for resistance to one of the rusts increases, identifying an unknown gene becomes more difficult. For example, suppose that a genetic study is done on a cultivar and resistance to stem rust proves to be due to a single gene. The pedigree of the cultivar may be useful in determining possible genes that it may carry, if the genotypes of some of its ancestors are known. In addition, the infection type conditioned by the gene to particular races may be helpful in its identification. Knott (1962a) made extensive use of this type of data to postulate probable genotypes for many closely related Kenyan cultivars.

Useful information can be obtained if cultivars or lines carrying single unidentified resistance genes are tested with a number of rust races. Since many cultivars carry several genes, the genes must first be isolated in separate lines. This can be done most easily by backcrossing the cultivar to a susceptible line, testing backcross F_2 families and identifying those that appear to be segregating for a single gene for resistance (i.e., a 3:1 or 1:3 ratio and often a single resistant infection type). Resistant plants in these families are selected and progeny tested. Those that prove to be homozygous should carry single genes. Single-gene lines can also be produced from a cross by selecting resistant plants from F_3 families which appear to be segregating for a single gene. When single gene lines have been produced, they can be tested with a number of races and their infection types compared with those of a set of lines carrying known genes for resistance. For example, near-isogenic lines were produced in a susceptible line, LMPG, for two genes for stem rust resistance, one derived from an einkorn wheat and one from Maribal/Marina. With nine races of stem rust, the line carrying a gene from einkorn gave a pattern identical to a line carrying *Sr9b* from Kenya 117A (Table 5.6). A cross between the two did not segregate, showing that the gene was indeed *Sr9b*. The line carrying a gene from Maribal/Marina gave an infection type 2 with all nine races, as did *Sr30* from Webster. Again a cross between the lines did not segregate.

Table 5.6. The infection types obtained when lines carrying *Sr9b, Sr30,* and two unidentified genes were tested with nine races of stem rust. (Knott unpublished data)

	Infection type with stem rust race								
	11	C15	15B-1	15B-1L	29	48A	56	111	C65
Sr9b/10*LMPG	2^+	4	4^-	2^-2	2^-2	2^-	2^-	2^-	2^-
Sr30/7*LMPG	2^-	2^-	2^-	2^-	2^-2	2^-	$2^=$	$2^=$	$2^=$
Einkorn/8*LMPG	2^-	4	4	2^-	2^-	2^-	2^-	$2^=2^-$	$2^=$
MM/8*LMPG	2^-	2^-	2^-2	2^-	2^-2	2^-	2^-	2^-	2^-

If none of these tests is successful in identifying an unknown gene, then two procedures are possible. Both require extensive crossing.

1. A line carrying the unknown gene can be crossed with as complete a set as possible of lines each carrying a single gene at a known locus for resistance. For each of the rusts a number of cultivars or lines carrying single genes for resistance to specific rust races are known. For both leaf and stem rust, sets of near-isogenic lines carrying individual genes at most of the known loci are available. The F_2 plants from a cross between two resistant lines must be tested with a race to which both are resistant. If the cross segregates susceptible plants, then different loci are involved. If a large sample (e.g., 400 or more) of F_2 plants is grown and all are resistant, then the genes are either allelic or linked. A very large sample is essential to distinguish between close linkage and allelism, since the F_2 generation is very inefficient in detecting linkage. For example, if both genes are dominant, the only detectable products of crossing over are the susceptible plants. Even with 10% crossing over, their frequency will be only 0.25%. A better test for linkage is to cross the F_1 plants to a susceptible line (i.e., a testcross).

Occasionally, no race will be available to which both genes in a cross give resistance. In such a case, the F_2 plants can be tested to two races, one avirulent on each gene for resistance. The plants can be inoculated with the first race and after about 7 or 8 days when pustules have appeared but not yet erupted, they can be inoculated with the second race. The first race is then read at about 14 days before the second race has developed sufficiently to confuse the readings. The leaves infected with the first race are removed. The second race is read at about 21 days. If some plants are susceptible to both races, then two loci are involved. If all plants are resistant to at least one race, then the genes are either allelic or closely linked.

When it appears that the two genes are either identical or allelic, then lines carrying the genes separately should be tested with a number of rust races. If they give identical infection types, the genes are probably identical.

If the unknown gene proves to be at a different locus than all those in the lines used as testers, then a monosomic analysis should be done to identify the chromosome carrying it (see Chap. 7).

2. An alternative way of studying an unknown gene for resistance is to begin with a monosomic analysis, crossing a line carrying the gene with a set of monosomic lines (see Chap. 7). In carrying out the rust tests, a race should be used to which the unknown gene is the only one giving resistance. This ensures that simple, single-gene ratios will be obtained, and simplifies the monosomic analysis. Once the chromosome carrying the unknown gene has been determined, crosses can be made to stocks carrying genes known to be on that chromosome. If the unknown gene proves to be at an identified locus, it may be necessary to test it with a number of races to determine whether it is a new allele. The advantage of this second procedure is that if the gene proves to be at a new locus, its chromosome location has been identified.

5.8 Genetic Linkages

Genetic linkages between genes for rust resistance and genes controlling other characters or resistance to other diseases can be useful in identifying the presence in unknown genotypes of particular genes for rust resistance. In breeding work they can sometimes be used as markers for the transfer of a specific gene for rust resistance. The presence of one gene may indicate the presence of a gene for resistance whose effect is masked by other genes for resistance. McIntosh and Cusick (1987) reviewed the known linkages in hexaploid wheat.

Frequently when a gene has been transferred to wheat from a related species, the transferred segment of chromatin carries several identifiable genes. If the chromatin is not homologous with wheat chromatin, then the genes will not be separated by crossing over and will be inherited together. For example, the 1BL-1RS translocation present in a number of wheats carries *Yr9, Lr26, Sr31* and *Pm8*. The presence of any one of the genes indicates the presence of all four. In at least some crosses, *Yr9* is recessive but *Lr26* is dominant. Heterozygotes can be identified with leaf rust but not with yellow rust. Similarly, in Agatha *Lr19* is linked to *Sr25* and a gene for yellow pigment, in Agent *Lr24* is linked to *Sr24* and in Transec *Lr25* is linked to *Pm7*.

Close linkages may also be useful although there is always a possibility that the two genes will be separated by crossing over. Some examples are *Yr7-Sr9g* (2BL-2% recombination), *Lr9-Sr11* (6BL - very low recombination), *Lr9*-B2 (awn inhibitor, 6BL - 4% recombination) and *Sr17-Pm5* (3–6% recombination).

In recent years increasing numbers of loci controlling enzymes and storage proteins in wheat have been identified. They have potential as genetic markers. For example, *Yr10* is linked to genes controlling glume color (*Rg1* - 2% recombination) and a storage protein (*Gli-B1*) on 1BS (Payne et al. 1986). The identification of restriction fragment length polymorphisms (RFLPs) in wheat will increase the genetic markers that are available to identify linkages of genes for rust resistance and other characters (Worland et al. 1987).

5.9 The Effect of Temperature on Genes for Resistance

Genes for resistance to all three rusts have been shown to be sensitive to temperature and occasionally to light. Most frequently as temperature increases resistance decreases (or susceptibility increases). For example, Bromfield (1961) tested 87 cultivars of bread and durum wheat with three races of stem rust at 70°F (21°C) and 85°F (29°C). With race 17, 50% of the cultivars were resistant at 70°F and susceptible at 85°F, with race 38, 31%, and with race 56, 37%. No cultivar was susceptible at 70°F and resistant at 85°F. However, 70°F is not a particularly low temperature and some of the cultivars that were resistant at both 70° and 85°F could have been susceptible at a lower temperature.

As more studies are done at lower temperatures, more examples are found of genes for rust resistance that are ineffective at low temperatures but effective at higher temperatures. Dyck and Johnson (1983) tested 27 near-isogenic lines of Thatcher carrying single genes for resistance to leaf rust. Tests were done with two isolates of leaf rust at 10, 15, 20 and 25°C, one isolate at 15 and 25°C and one isolate at 15 and 20°C. The *Lr18* line and the cultivar Frontana showed fairly large increases in susceptibility at higher temperatures. Several other lines showed smaller increases. However, the *Lr3, Lr16, Lr17,* and *Lr23* lines showed large increases in resistance at higher temperatures, and several other lines showed smaller increases. In some cases the changes caused by temperature occurred with one isolate but not with another. This suggests that other genes in the pathogen can modify the temperature sensitivity of the interaction between a specific gene for resistance in the host and the corresponding gene for resistance in the pathogen.

Browder and Eversmeyer (1986) studied near-isogenic lines of Thatcher carrying *Lr1, Lr16,* or *Lr17.* The lines were inoculated with an avirulent culture, held at 16°C overnight, then held at 20°C for 0, 2, 4 or 8 days before being transferred to 5°C. In the first environment (0 days at 20°C), all three lines were susceptible. Two days at 20°C were sufficient to induce resistance in the *Lr1* line, 4 days were required for the *Lr17* line and 8 days for the *Lr16* line.

The gene *Sr6* has been studied extensively for sensitivity to light and temperature. Forsyth (1956) found that the resistance conditioned by *Sr6* broke down between 20 and 22°C. Seedlings carrying *Sr6* were susceptible if grown under a regime of 16 h of light at 24°C and 8 h of darkness at 16°C. However, if the seedlings were given a 16 h light period at 16°C between 52 and 91 h after inoculation, they were resistant. Thus, there is a critical interaction between light and temperature.

Luig and Rajaram (1972) tested seedlings carrying *Sr6* under continuous light at temperatures from 15°C to 30°C, using 3°C intervals. Resistance began to break down at 18°C and had disappeared by 24°C. Antonelli and Daly (1966) carried out tests in which seedlings carrying *Sr6* were held at 20°C for various times after inoculation and then transferred to 25°C, or vice versa. If they were held at 20°C for 3 days and then transferred to 25°C, they were susceptible. If they were transferred at 5 or 7 days, they showed some evidence of resistance and if they were transferred at 9 days they showed nearly normal resistance. In the reciprocal test, when seedlings were held at 25°C for 3 days after inoculation and then transferred to 20°C, they were resistant. However, if they were held at 25°C for 4 days, resistance began to break down. Both rests indicated that the temperature for the first 3 days after inoculation was most critical. Knott (1981) studied *Sr6* in crosses between Kenya 58/10* Marquis (*Sr6*) and four susceptible lines or cultivars. The sensitivity of heterozygotes to temperature and the dominance of *Sr6* varied in different genetic backgrounds. Thus, the resistance conditioned by *Sr6* is determined by a complex interaction between genotype and environment.

Clearly the rusts interacts strongly with their environment, particularly temperature, light and, of course, moisture.

5.10 The Inheritance of Complex Resistance

Ausemus et al. (1946) listed a number of papers which reported that resistance to leaf or stem rust of wheat was multigenic. Similarly, Person and Sidhu (1971) surveyed 301 papers dealing with rust resistance and found that 19 reported the occurrence of minor genes or polygenes.

With the increased interest in horizontal or general resistance, adult plant resistance, slow rusting, tolerance, etc., stimulated by Vanderplank (1963), many more cases of complex resistance have been reported. Since these types of resistance are more complex in inheritance, they are more difficult to analyze genetically. As described in Chap. 3, special methods are often necessary to measure them, e.g., measurement of the latent period, pustule size, spore production, or area under the disease progress curve.

Genes for specific resistance can often mask genes controlling more complex resistance. Either the genes for specific resistance have to be removed in some way or a race used that is virulent on any genes for specific resistance that are present in the crosses being studied.

Many of the studies on the inheritance of complex resistance have used standard procedures, the testing of F_1 and F_2 plants and F_3 families, and occasionally backcrosses. Complex resistance to yellow rust of wheat has frequently been reported (reviewed by Robbelen and Sharp 1978). Typically, the number of resistant progeny in the early generations of crosses is low, there are many intermediate types of reaction, and there is no distinct separation between resistant and susceptible plants. Resistance is often affected by environment, particularly temperature. Lewellen et al. (1967) and Lewellen and Sharp (1968) studied F_1, F_2, F_3 and testcross populations from several crosses at two temperature profiles, 2°C night/18°C day and 15°C night/24°C day, in a growth chamber. The results indicated that resistance was controlled by two types of genes. In addition to dominant major genes (in one case complementary dominant genes), there were several recessive minor genes. In Rego the minor genes were effective at lower temperatures, but in P.I. 178383 they were effective at higher temperatures. The minor genes from the two cultivars could be combined to give good resistance at both temperature profiles.

From the cross, Itana (susceptible) X P.I. 178383, Sharp and Volin (1970) selected lines carrying 1, 2 or 3 minor genes. At 2°C/18°C they gave IT's 3, 2, and 0, respectively, on a 0–4 scale, and at 15°C/24°C they gave IT's 2, 0, and 0^-. The lines gave similar reactions to 11 isolates of stripe rust, and the authors suggested that the resistance was nonspecific.

Krupinsky and Sharp (1978) studied the parents and F_1 plants from a 6×6 set of diallel crosses involving six minor gene lines, and from a 7×7 set of diallel crosses involving three minor gene lines and four susceptible winter wheats. They concluded that general combining ability (additive gene effects) was much larger than specific combining ability, and heritabilities were above 90%. Maternal and reciprocal effects were also significant.

Pope (1968) did field studies on numerous crosses and indicated that resistance to yellow rust was frequently due to a minimum of five genes having small, additive effects. He suggested that a minimum of 20 genes was involved and that they functioned in gene complexes.

In England a number of the older wheat cultivars are susceptible to yellow rust as seedlings but have maintained adequate resistance in the field. Lupton and Johnson (1970) crossed the resistant Little Joss with the susceptible Nord Desprez. The F_1 plants were severely rusted and only 5% of 2500 F_2 plants showed some field resistance. Their F_3 progenies showed varying degrees of resistance. The authors concluded that resistance was recessive and complex in inheritance. Wallwork and Johnson (1984) intercrossed Maris Bilbo, Joss Cambier, and Nord Desprez which are susceptible to race 104 E137(3) in the field. Transgressive segregation for resistance occurred in all three crosses. Three of the most resistant lines from Joss Cambier x Nord Desprez showed more resistance than their parents to 12 races. The authors suggested that resistance was recessive and that the number of genes was not large since resistance was accumulated quickly.

In recent years there has been increasing interest in resistance that reduces the rate of epidemic development by reducing the number and rate of growth of pustules, the final pustule size and the spore production (reviewed by Parlevliet 1979, 1983, 1985). The effect is often called slow rusting and the resistance is thought to be nonspecific. However, slow rusting is simply one manifestation of resistance and is not necessarily nonspecific. Specific genes conditioning intermediate levels of resistance also reduce the rate of epidemic development because of reduced spore production (and possibly longer latent periods as well).

Skovmand et al. (1978b) studied the inheritance of slow rusting to stem rust in a set of diallel crosses involving seven parents. The parents varied in their speed of rusting and in the genes for specific resistance that they carried. Random F_5 lines developed by single seed descent were evaluated for 2 years in the field. Transgressive segregation for resistance occurred even in the cross involving the two fastest rusting parents. From 2 to 12 genes were estimated to be segregating in different crosses. Although the estimates are probably subject to considerable error, the results suggest that a number of genes with small, cumulative effects are involved. Skovmand et al. (1978a) measured the effect of specific *Sr* genes in the same set of crosses. The *Sr* genes had effects of various sizes, probably depending on the frequency of races to which they gave resistance (although this is not discussed by the authors). The authors suggested that the two genes with the largest effects, *Sr6* and *SrTt1*, were not themselves causing the effects, but were linked to genes causing slow rusting. However, Rowell (1981) found that *SrTt* is associated with slow rusting and low receptivity, and concluded that resistance may be a pleiotropic effect of a single gene. Cox and Wilcoxson (1982) concluded that *Sr6* acts in concert with other unidentified genes. However, Johnson (1984) suggested that *Sr6* provided resistance to most of the races present in their experiments and could, therefore, have caused slow rusting by its direct effect in a heterogeneous rust population.

Martinez-Gonzalez et al. (1983) studied two crosses involving Era which carries *Sr5, 6, 8, 9a* or *9b, 11, 12,* and *17*. Lines which carried *Sr6* and were resistant to race 15B2-TLM were eliminated and susceptible F_5 lines were tested for 2 years in the field (a different set of lines each year). They concluded that slow rusting was a quantitative character.

Martin et al. (1977) studied six cultivars with stem rust race 15B-2. They found no correlation between field reaction to rust and percent penetration. Slow rusters showed slower hyphal growth, fewer haustoria, and more necrotic cells. Martin et al. (1979) found that the speed of rusting for five lines in the field did not always agree well with the size and number of uredia on seedlings or adult plants in the greenhouse.

Rather similar results have been obtained with leaf rust. Ohm and Shaner (1976) found that slow rusters differed from fast rusters in length of the latent period, pustule size, and pustule numbers, but not necessarily in all three in any one line. Kuhn et al. (1980) studied 81 F_3 families from the cross between a slow ruster, Suwon 85, and a fast ruster, Suwon 92. They concluded that a short latent period was partially dominant and that two genes were involved. However, the great variability in the parents makes the results dubious.

Knott (1982) developed lines of wheat that lacked seedling genes for resistance to race 15B-1 of stem rust but had good field resistance to the same race. The F_2 seedlings from crosses involving four selected parents were tested with race 15B-1 in the greenhouse and resistant plants eliminated. The progeny of the susceptible plants were then selected for several generations for resistance to race 15B-1 in artificially inoculated field nurseries. Lines with low rust severities were readily obtained. They proved to have resistance to a broad range of North American races and resistance appeared to be complex in inheritance.

The extensive data that are now available indicate clearly that there is a type of resistance to rust that is controlled by several genes having small, additive effects. Although tests have been limited, there is some evidence that resistance of this type is durable. Parlevliet (1983) argues that some race-specific genes may have effects that are too small to be recognized and are hidden by the experimental error. The resistance is then said to be horizontal. Parlevliet (1983) also suggests that durability does not necessarily depend on race nonspecificity, and that the two characters are different and sometimes independent. Unfortunately, if some genes have effects that are so small that they are hidden by the experimental error, then the hypothesis that they are race-specific cannot be tested.

5.11 Procedures for Studying Complex Resistance

Resistance that is complex in inheritance is difficult to study genetically. Individual genes have small effects and distributions in segregating generations of crosses is usually continuous. Thus, normal genetic analyses using F_2 populations, F_3 families, and backcrosses are more difficult than for simply inherited resistance. If resistance is measured on an appropriate scale, such as the percent rust intensity in field tests, then quantitative genetic methods can be used. Depending on the experimental design, additive, dominance and epistatic effects, heritability, and the number of effective factors can be estimated. Particularly for the number of effective factors, the assumptions made in the analyses are often unrealistic and the estimates are subject to considerable error.

An alternative way of studying complex resistance is to test a set of random lines developed from a cross by single seed descent (SSD). The resistant line to be analyzed is crossed to a susceptible parent. Starting with the F_2 population, one seed is taken from each plant and grown to produce the next generation. At the F_5 or F_6 generation when the plants are reasonably homozygous, individual plant progenies can be grown to produce seed for rust testing. Since in most cases some plants will be lost in each generation due to failure of germination or other causes, it is necessary to start with more F_2 plants than the number of lines desired. One hundred lines is probably a minimum with more lines being required the more complex the inheritance. The loss of plants during single seed descent can be minimized by sowing two seeds per plant and thinning to one seedling. However, this requires more labor.

If a single locus is involved, then F_5 plants produced by SSD should be 15/32 (0.469) homozygous resistant, 15/32 homozygous susceptible and 2/32 (0.062) segregating. Normally, F_6 lines derived from individual F_5 plants would be tested for rust reaction in a field nursery. The proportion of F_5 plants homozygous for all segregating genes for resistance can be calculated for any number of loci: 2 genes – 0.220, 3 genes – 0.103, 4 genes – 0.048, 5 genes – 0.023, etc. These figures are the proportion of F_5 plants (or F_6 lines) that should be as resistant as the resistant parent. If several genes having small, cumulative effects are involved, then the distribution of the F_6 lines will probably be continuous. For example, if five loci are segregating, then lines that are homozygous resistant at four loci but segregating at one may not be distinguishable from lines homozygous at all five. Nevertheless, tests on such a set of lines can give an indication of the number of genes involved.

As noted earlier, Knott (1982) produced a number of lines that lacked seedling genes for resistance to race 15B-1 of stem rust but had good field resistance to 15B-1 and many other races. Knott and Padidam (1988) studied F_5-derived F_7 lines from crosses between six of these parents and a susceptible line. In each cross, the F_7 lines gave distributions that were strongly skewed to the resistant side (i.e., susceptible

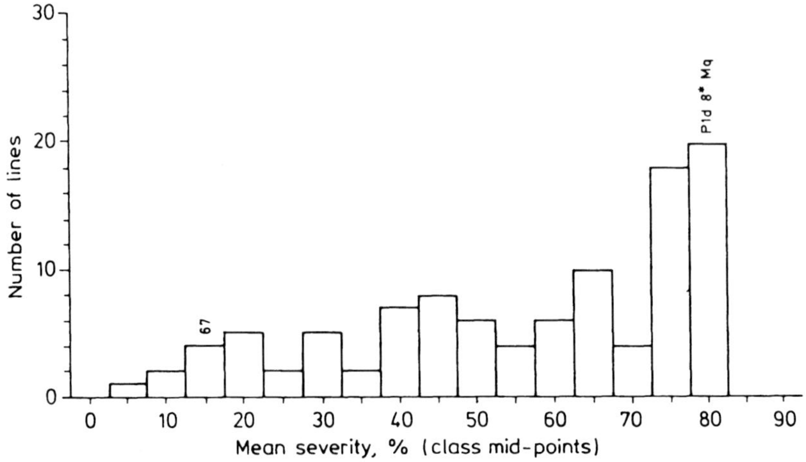

Fig. 5.1. Distribution of mean rust severities of 104 F_7 lines from a cross between line 67 and a susceptible parent, Pld 8*Mq, tested in a field rust nursery with race 15B-1 in 1982

lines were much more frequent than resistant lines) (Fig. 5.1). The analyses indicated that at least three genes were involved in each parent. The data suggested that lines that were homozygous for only one gene were little different than the susceptible parent. Apparently two or more genes had to be combined to produce significant resistance, and the gene effects appeared to be multiplicative rather than additive.

5.12 The Inheritance of Virulence in Rust Fungi

Craigie (1927, 1931) discovered the sexual cycle in the rust fungi. The research group at the former Dominion Rust Research Laboratory, Winnipeg, Manitoba, then studied the inheritance of spore color and pathogenicity in stem rust. In an early study, Newton et al. (1930b) had difficulty interpreting the inheritance of virulence. However, if virulence on individual cultivars is examined, their data indicated that virulence was due to a single recessive gene in one case, a single dominant gene in another, and three recessive genes in a third. They also reported some cytoplasmic effects, although the differences between reciprocal crosses were relatively small (ITs 3^- and 4). This work was followed by a number of studies such as those by Luig and Watson (1961), Loegering and Powers (1962), Green (1966), and Kao and Knott (1969), which established the basic principles for the inheritance of virulence in stem rust. In general, virulence is due to single, recessive genes. However, virulence on a gene in Mindum and Arnautka is clearly dominant.

Several studies on the inheritance of virulence in leaf rust have been carried out, particularly at the Agriculture Canada Research Station in Winnipeg, and at North Dakota State University. In general, virulence has been due to single, recessive genes (Haggag et al. 1973; Samborski and Dyck 1968, 1976; Statler 1977, 1979; Statler and Jones 1981).

Samborski (1963) discovered a single leaf rust pustule of IT 1⁺ on Transfer, which normally gives IT 0. An isolate was produced from the pustule and selfed. It proved to be heterozygous for virulence on Transfer (*Lr9*). The various combinations of the genes at the *Lr9* and *p9* loci resulted in the following infections types:

	Lr9 Lr9	*Lr9 lr9*	*lr9 lr9*
P9 P9	0;	0;	4
P9 p9	1^+	3	4
p9 p9	4	4	4

Thus, virulence was partially dominant, particularly on the *Lr9 lr9* heterozygote. Virulence in the leaf rust fungus appears to have more tendency to be dominant than does virulence in stem rust.

CHAPTER 6
The Genetics of Host-Pathogen Interactions

6.1 Introduction

A rust disease involves a very intimate interaction between the host and an obligate pathogen. The result of the interaction can lead to differential selection in either the host or the pathogen. A very high level of virulence can prevent the reproduction of a wheat plant while a very high level of resistance can prevent the survival of the rust fungus. It is not surprising, therefore, that very close genetic relationships have developed between wheat and its rusts.

6.2 The Gene-for-Gene System

6.2.1 Development of the Concept

Flor (1942, 1946, 1947) was the first person to study both the inheritance of pathogenicity in a pathogen and the inheritance of reaction in its host. Working with flax rust (*Melampsora lini* Desm.) and its host, flax (*Linum usitatissimum* L.), he demonstrated that if a flax cultivar carried a single gene for resistance, then virulence in the rust was also conditioned by a single gene. Similarly, if resistance in a cultivar was conditioned by two genes, then virulence in the rust was conditioned by two genes. In 1942, Flor stated: "these facts suggest that the pathogenic range of each physiologic race of the pathogen is conditioned by pairs of factors that are specific for each different resistant or immune factor possessed by the host variety." Later Flor (1956) used the term complementary genes for the interacting genes in the host and pathogen. However, since this term has a specific and different genetic meaning, the terms corresponding or matching genes are now used.

The significance of Flor's work was largely overlooked until Person (1959) published a theoretical analysis of gene-for-gene relationships in host:parasite systems. He concluded that such relationships should occur as a general rule in host:parasite systems as a result of selection pressures during evolution. Person used the broader term parasite, but this book concerns plant pathogens and, therefore, the term pathogen will be used.

Gene-for-gene systems are thought to have developed by a series of evolutionary steps in nature (Table 6.1). If a host is initially susceptible to a pathogen, then a resistant mutant will have a selective advantage when the host is being attacked. Natural selection will favor an increase in the frequency of the mutant gene in the host. As the gene for resistance increases in the host population, a virulent mutant in the pathogen that can overcome the gene for resistance in the host will have a

Table 6.1. Steps in the evolutionary development of two pairs of corresponding genes in a host-pathogen system in nature

	Initial situation	Step 1	Step 2	Step 3	Step 4
Host genotype	*r1r1r2r2* →	*R1R1r2r2*	*R1R1r2r2*[a] →	*R1R1R2R2*	*R1R1R2R2*
Pathogen genotype	*P1P1P2P2*	*P1P1P2P2* →	*p1p1P2P2*	*p1p1P2P2* →	*p1p1p2p2*
Interaction phenotype	HIT[b]	LIT[b]	HIT	LIT	HIT
Selection pressure	On host	On pathogen	On host	On pathogen	On host

[a] At this point there are two possibilities, *R1* may mutate to a new allele for resistance or *r2* may mutate to *R2*. Only the latter possibility is shown.
[b] HIT = high infection type and LIT = low infection type

selective advantage. In fact, on an immune host plant only the mutant can survive. Once the virulent mutant has become common, selective advantage will again favor a resistant mutant in the host and the process will repeat itself. This time two types of mutants are possible in the host: (1) the first gene for resistance may mutate to a new allele that is not overcome by the first gene for virulence in the pathogen, or (2) a mutation to resistance may occur at a new locus. In either case, as the mutant gene for resistance increases in frequency, selection will favor a new gene for virulence in the pathogen.

Person et al. (1962) defined the gene-for-gene concept as follows: "A gene-for-gene relationship exists when the presence of a gene in one population is contingent on the continued presence of a gene in another population, and where the interaction between the two genes leads to a single phenotypic expression by which the presence or absence of the relevant gene in either organism may be recognized." For a wheat rust, a gene for resistance has no selective advantage unless the pathogen carries the corresponding gene for avirulence, and a gene for virulence has no selective advantage if the host does not carry the corresponding gene for resistance. It is assumed in each case that the gene for resistance or for virulence has no other function that is advantageous.

Although there are exceptions, many genetic analyses have shown that resistance to rusts in wheat is controlled by single dominant genes and virulence in the pathogens is controlled by corresponding recessive genes (or to put it the other way, avirulence is controlled by corresponding dominant genes). The fact that resistance and avirulence are generally dominant, suggests that they result from the production of active gene products. In those cases where they are recessive, dosage effects may be involved. A single dose of the gene may not result in the production of sufficient active gene product to reach a threshold level, but two doses do.

A commonly held theory is that incompatibility (resistance) is a recognition phenomenon involving an active gene product from a resistant host and an active gene product from an avirulent pathogen. During evolution a recessive gene for susceptibility in a host can mutate to a dominant gene for resistance. The latter must produce a gene product which interacts with a gene product produced by an existing gene in the pathogen. Mutations from a recessive gene to a dominant gene producing an active product are likely to be rare. On the other hand, since virulence is generally recessive, a mutation in the pathogen from a dominant gene for avirulence to a recessive gene that results in a product that no longer interacts with the product of the gene for resistance, will result in virulence. The occurrence of allelic series at several loci for resistance to a rust in wheat suggests that one allele for resistance can mutate to a new allele for resistance which no longer interacts with the gene for avirulence corresponding to the first allele. Less work has been done on the inheritance of virulence in the pathogen. However, Kao and Knott (1969) did find that different loci controlled virulence on *Sr9a* and *Sr9b*, i.e., the two alleles are interacting with different loci in the rust. The situation seems to be more complicated at the *Lr2* locus at which three alleles are known (Dyck and Samborski 1974). Virulence is controlled by a single recessive gene, *p2*, but an additional gene modifies or inhibits the action of *p2*.

The infection type produced when a pathogen attacks a host plant is the product of the interaction of two genetic systems. For a single pair of corresponding genes, if it is assumed that only homozygotes occur, the interaction can be as illustrated in Table 6.2. Only one combination, *RR/PP*, results in a low infection type (LIT) – the host carries a gene for resistance for which the pathogen does not carry the corresponding gene for virulence. Thus, specificity occurs in the combination involving the genes that are normally dominant, *R* and *P*. Any one of the three other combinations, *RR/pp*, *rr/PP* or *rr/pp*, results in susceptibility.

Table 6.2. The interaction between genotypes at one pair of coresponding genes

		Pathogen genotype	
		PP	*pp*
Host	*RR*	LIT[a]	HIT[a]
Genotype	*rr*	HIT	HIT

[a] LIT = low infection type, HIT = high infection type

Contrary to the general opinion, Vanderplank (1982, 1984) argued that susceptibility is specific because a plant with a gene for resistance *R1* is susceptible only if the rust carries *p1*. This overlooks the fact that two other combinations, *r1r1/P1P1* and *r1r1/p1p1*, also result in susceptibility whereas only one combination results in resistance. He also stated that since a plant with *R1* is resistant to rust isolates carrying any other genes for virulence such as *p2, p3, p4*, and *p5*, resistance is "unspecific". This, of course, misses the whole point of the gene-for-gene hypothesis – the genetic and the physiological interactions involve only corresponding genes in the host and pathogen, and their gene products. Only one specific combination (*R1R1/P1P1*) of the genes in a corresponding gene pair results in a LIT.

A theoretical gene-for-gene system with three corresponding pairs of loci is given in Table 6.3. Person (1959) studied such a system and described its characteristics, including the following:

1. If there are n pairs of corresponding loci and only two phenotypes for each locus, then there are 2^n possible host phenotypes and 2^n possible pathogen phenotypes.

Table 6.3. A theoretical gene-for-gene system with three corresponding pairs of loci in the host and pathogen. (After Person 1959)

Pathogen isolates and their genes for virulence	Host lines and their genes for resistance								No. of hosts attacked	Loci for virulence
	H1 --	H2 *R1*	H3 *R2*	H4 *R3*	H5 *R1R2*	H6 *R1R3*	H7 *R2R3*	H8 *R1R2R2*		
P1 --	H[a]	–[b]	–	–	–	–	–	–	1	0
P2 *p1*	H	H	–	–	–	–	–	–	2	1
P3 *p2*	H	–	H	–	–	–	–	–	2	1
P4 *p3*	H	–	–	H	–	–	–	–	2	1
P5 *p1p2*	H	H	H	–	H	–	–	–	4	2
P6 *p1p3*	H	H	–	H	–	H	–	–	4	2
P7 *p2p3*	H	–	H	H	–	–	H	–	4	2
P8 *p1p2p3*	H	H	H	H	H	H	H	H	8	3
No. of virulent isolates	8	4	4	4	2	2	2	1		
Loci for resistance	0	1	1	1	2	2	2	3		

[a] H = a high infection type
[b] – = a low infection type

2. One host genotype (H1), the universal suscept, is susceptible to all races, and one genotype (H8) is susceptible to only one race of the pathogen, the one having universal virulence.

3. One pathogen isolate (P8) is virulent on all host genotypes and one (P1) is virulent on only one, the universal suscept.

4. The host genotypes fall into groups that form a geometric series according to the number of races that can attack them (1, 2, 4, 8, 16, 32, etc.).

5. The pathogen races fall into a similar series according to the numbers of genotypes they can attack.

6.2.2 Evolution of Gene-for-Gene Systems in the Field

Under commercial wheat production, wheat breeders and farmers determine the evolution of the host. The wheat breeders produce new cultivars with new combinations of genes for rust resistance and the farmers determine which will be grown. The effect of the resulting directional selection on the pathogen has been demonstrated in many countries. When a previously unused resistance gene is released in a cultivar, the corresponding virulence usually appears in the field sooner or later.

Perhaps the best example of changes that can occur in a rust pathogen has come from Australia, where the evolution of wheat stem rust races has occurred very rapidly for many years. Watson and Luig (1963) documented some of these changes (Table 6.4). In each case when a new gene for resistance was introduced, it was rapidly overcome by the pathogen. This led the Australian wheat breeders to use combinations of genes which were more difficult for the pathogen to overcome.

Table 6.4. Some examples of changes in Australian races of stem rust in response to the release of new, rust-resistant cultivars. (After Watson and Luig 1963)

Year	Cultivar and genotype[a]	Race present	Virulence
1938	Eureka - *Sr6*	126-6	Avirulent on Eureka
1942		126-1,6	Virulent on Eureka
1945	Gabo types - *Sr11*	222-?	Avirulent on Gabo
1948		222-2,6 and 222-1,2,6	Virulent on Gabo
1950	Festival - *Sr9*	21-2	Avirulent on Festival
1959		21-2,3	Virulent on Festival
1958	Mengavi - *SrTt1*	34-2	Avirulent on Mengavi
1960		34-2,4	Virulent on Mengavi

[a] The new gene for stem rust resistance that was introduced

In North America the situation is somewhat more complex because a wide range of both winter and spring wheats are grown over a large part of the continent. The cultivars vary widely in the degree of resistance they have and the genes for resistance they carry. As a result, directional selection on the rust pathogens is not as great as it would be if only a few resistant genotypes were present. Nevertheless, the effect of new cultivars on the genes for virulence in leaf and stem rust has been clear in a number of cases. Green (1975) showed that changes in stem rust races often occurred by changes in virulence on one host gene at a time (Fig. 6.1). Usually the changes involved the gain of a gene for virulence but occasionally the loss of a gene occurred. In two cases the changes starting with C9 appeared to involve two loci at once. It is possible, however, that single changes occurred but the genotype resulting from the intermediate step was never picked up in the race surveys in Canada.

The experience with all three wheat rusts is that when a cultivar carrying a gene for resistance is released, the gene is frequently matched by the corresponding gene for virulence. Virulence is known for almost every resistance gene that has been identified, but fortunately combinations of resistance genes may remain effective for long periods. One possible exception is the gene *Sr26* for stem rust resistance derived from *Agropyron elongatum* (Host) Beauv. (*Elytrigia pontica* (Podp.) Holub) (Knott 1961). In a world-wide survey of virulence genes, Luig (1983) found that cultivars carrying *Sr26* were resistant to all cultures of stem rust. Although Luig (1979) reported that no mutants virulent on *Sr26* were produced using ethyl methane sulphonate (EMS), a virulent isolate was later obtained (Luig personal communication 1984). However, the isolate proved to be a very weak competitor and this may account for the fact that it has not been found in the field. Since *Sr26* is on an alien chromosome segment within which crossing over does not occur, it is just possible that it is a gene complex rather than a single gene.

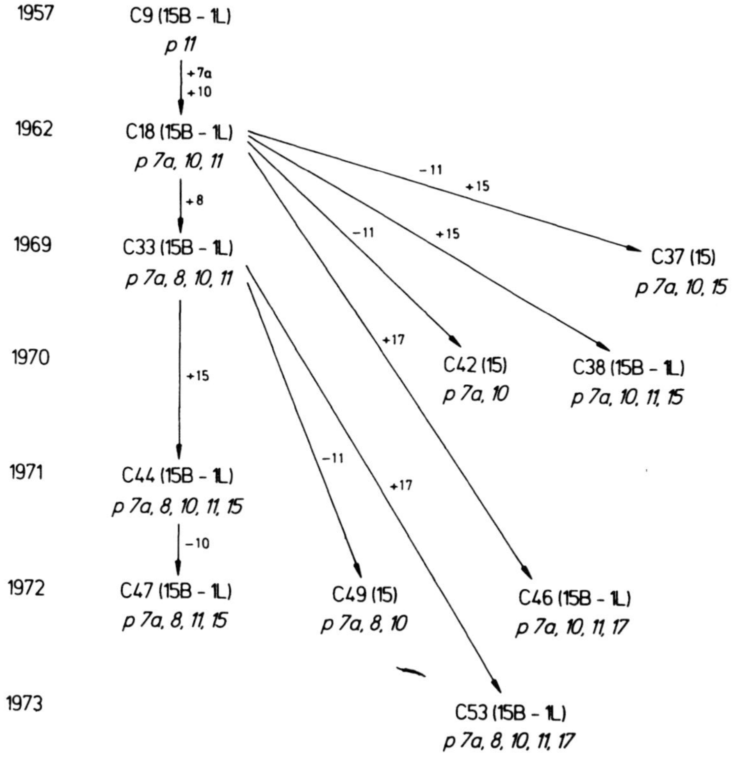

Fig. 6.1. The evolution of some stem rust races in North America, usually involving changes in virulence at one locus at a time. (After Green 1975)

6.2.3 Gene-for-Gene Systems in the Rusts

In the years following Person's paper (1959), gene-for-gene relationships were demonstrated or hypothesized in a number of host-pathogen systems including the rusts (Day 1974). Luig and Watson (1961) studied the inheritance of pathogenicity in *Puccinia graminis* f. sp. *tritici* and showed that in some cases genes for virulence matched known genes for resistance in the host. Loegering and Powers (1962) also studied pathogenicity in *P.g. tritici* and used the gene-for-gene hypothesis to postulate genotypes for resistance in wheat cultivars. Green (1964, 1966) showed that in *P.g. tritici* virulence on six lines of Marquis carrying single resistance genes was conditioned by single, recessive virulence genes. N.D. Williams et al. (1966) found that pathogenicity on lines carrying single genes for resistance from Marquis and Reliance was controlled by single genes in the rust. Kao and Knott (1969) crossed races 111 and 29 of stem rust and found that virulence on six *Sr* genes was controlled by a single, recessive gene for virulence in each case. There was some evidence that virulence on *Sr1* (now *Sr9d*) was due to two complementary genes but this was never confirmed.

Dyck and Samborski (1968a) studied the inheritance of resistance to *Puccinia recondita* in a number of the standard differential cultivars and identified a number of genes for resistance. Samborski and Dyck (1968) studied the inheritance of virulence on several of the same genes and concluded that the results were compatible with the gene-for-gene theory. Samborski and Dyck (1976) and Statler (1977, 1979) selfed and crossed several leaf rust races and studied the inheritance of virulence. Virulence was usually controlled by single, recessive genes but in several cases single, dominant genes were reported. Several pairs of genes for virulence were closely linked.

Since a sexual cycle has not been discovered in *Puccinia striiformis*, a direct proof of the occurrence of a gene-for-gene system cannot be obtained. However, such a system has been hypothesized by Zadoks (1961), Lewellen et al. (1967), and Line et al. (1970).

6.2.4 Interactions Between Host and Pathogen

Loegering and Powers (1962) emphasized the need for a standard set of terms to describe host-pathogen interactions. They proposed the following: the character of the host is its reaction, which may be resistant or susceptible; the character of the pathogen is its pathogenicity, which may be virulent or avirulent; and the interaction results in an infection type which may be low or high. These terms are now generally used.

For gene-for-gene systems, Loegering and Powers (1962) and Loegering (1966) then defined four categories of genetic relationships that determine the infection type:

Category I – Relationships between alleles at one locus in one organism (i.e., dominance or recessiveness of *R* and *r*, and of *P* and *p*). As already indicated resistance is often dominant and virulence recessive.

Category II – Relationships between genotypes at two or more loci in one organism (independent action or various types of interaction between genes at different loci in one organism). Resistance genes in the host frequently act independently of one another. Thus, if a cultivar carries resistant genes at two loci it will be resistant to all races to which either or both genes provide resistance. If both genes condition resistance but different infection types result, then the lower infection type is commonly expressed. The gene conditioning the lower infection type is often said to be epistatic over the second gene. While it is true that the effect of the second gene is masked, it is still active, as can be shown by testing with a race to which only the second gene conditions resistance, so this is not a true epistasis. Various types of interactions between resistance genes such as additive effects and complementary action are occasionally found.

Less work has been done on the rust pathogens but again the genes for avirulence frequently act independently of each other, although interactions have been reported.

Category III – Relationships between genotypes at one pair of corresponding loci in the host and pathogen. Since there are three possible genotypes at a locus in each of the host and pathogen, there are nine possible combinations (Table 6.5). The

Table 6.5. Infection types for the nine combinations of wheat and rust genotypes for one pair of corresponding loci, depending on whether resistance and avirulence are dominant or recessive

	Host	Pathogen genotype					
	genotype	*PP*	*Pp*	*pp*	*PP*	*Pp*	*pp*
		Avirulence dominant			Avirulence recessive		
Resistance dominant	*RR*	LIT[a]	LIT	HIT[a]	HIT	HIT	LIT
	Rr	LIT	LIT	HIT	HIT	HIT	LIT
	rr	HIT	HIT	HIT	HIT	HIT	HIT
Resistance recessive	*RR*	HIT	HIT	HIT	HIT	HIT	HIT
	Rr	HIT	HIT	HIT	HIT	HIT	HIT
	rr	LIT	LIT	HIT	HIT	HIT	LIT

[a] LIT = low infection type and HIT = high infection type.

infection types produced will depend on the dominance of the genes for reaction and pathogenicity. The standard situation is shown in the upper left section of the table, where resistance and avirulence are both dominant. Regardless of dominance, a low infection type results only when the host carries a gene for resistance and the pathogen carries the corresponding gene for avirulence.

Category IV – Relationships involving two corresponding gene pairs. If the infection type for a particular host:pathogen combination is controlled by two corresponding gene pairs then the lower of the two LITs will be expressed, i.e., if one corresponding gene pair conditions an infection type 0 and another an infection type 2, the result will be type 0. A high infection type results only if all genes for resistance in the host are matched by the corresponding genes for virulence in the pathogen. Conversely, a low infection type will result as long as there is a single gene for resistance which is not matched by the corresponding gene for virulence. Three examples are given below, assuming that resistance and avirulence are dominant:

Host	*R1 R2 R3 r4*	*R1 R2 R3 r4*	*R1 R2 R3 r4*
Pathogen	*P1 P2 P3 P4*	*p1 p2 P3 P4*	*p1 p2 p3 P4*
Infection type	0 1 2 4	4 4 2 4	4 4 4 4
	= *0*	= *2*	= *4*

6.3 Using Infection Type Data Sets to Postulate Genotypes

Loegering et al. (1971), Loegering and Burton (1974) and Loegering (1984) developed a computer-based system for analyzing sets of infection type data and postulating genotypes of the host and pathogen. As noted previously, for a single pair of corresponding genes with resistance dominant and virulence recessive, there are four possible combinations – *RR*/*PP* = LIT, *RR*/*pp* = HIT, *rr*/*PP* = HIT, and *rr*/*pp* = HIT. Whenever a LIT is obtained, the host/pathogen genotypes must be *RR*/*PP*. Loegering (1984) refers to a LIT as the definitive phenotype. When a HIT is obtained, it can result from any one of three host/pathogen genotype combinations. However, a HIT does mean that either the host is *rr* or the pathogen is *pp*. If the pathogen is known to be *PP* then the host must be *rr* in order to obtain a HIT. Similarly, if the host is known to be *RR*, then the pathogen must be *pp* to result in a HIT.

The analytical procedure involves looking at the infection types produced by pairs of rust isolates on pairs of wheat cultivars. The simplest case is shown below:

1\.

	Isolate		Postulated genotypes		
	I1	I2		I1 - *P1P1*	I2 - *p1p1*
Cultivar C1	LIT	HIT	C1 - *R1R1*	LIT1	HIT
C2	HIT	HIT	C2 - *r1r1*	HIT	HIT

Since isolate I1 on cultivar C1 results in a LIT, which is designated as LIT1, C1 must carry a gene for resistance, *R1*, and I1 must carry the corresponding gene for avirulence, *P1*. Now, since I2 combined with C1(*R1R1*) results in a HIT, I2 must carry *p1*, the corresponding gene for virulence on *R1*. Since I1 (*P1P1*) combined with C2 results in a HIT, C2 cannot carry *R1* and must be *r1*. Thus, the genotypes of both cultivars and both isolates can be determined.

Seven basic boxes can be obtained by inoculating two wheat cultivars with two rust isolates, and varying amounts of genetic information can be obtained from each. Note that in the simplest box with one LIT and three HIT's, the LIT can appear in any one of the four corners, but this does not change the basic analysis. In fact, by changing the position of the isolates and cultivars, the box can always be rearranged so that the LIT is in the upper left hand corner. The remaining six boxes and the genotypes postulated from them are as follows:

2\.

	I1	I2		*P1P1*	??
C1	LIT	LIT	*R1R1*	LIT1	LIT?
C2	HIT	HIT	*r1r1*	HIT	HIT

In box 2, three genotypes can be determined but not the genotype of isolate 2. Isolate 2 could be *P1P1* in which case LIT? is another LIT1 (i.e., it results from the combination of *R1R1*/*P1P1*). If isolate 2 is not *P1P1*, then LIT? results from the interaction of a second corresponding gene pair (CGP).

3\.

	I1	I2		*P1P1*	*p1p1*
C1	LIT	HIT	*R1R1*	LIT1	HIT
C2	LIT	HIT	??	LIT?	HIT

Box 3 is similar to box 2 but this time it is cultivar 2 whose genotype is unknown. The LIT? could be caused either by the first CGP or by a second pair.

4\.

	I1	I2		*P1P1p2p2*	*p1p1P2P2*
C1	LIT	HIT	*R1R1r2r2*	LIT1	HIT
C2	HIT	LIT	*r1r1R2R2*	HIT	LIT2

Box 4 permits the complete identification of the genotypes for two CGPs. The two LITs are in diagonally opposite corners and must result from two different *RR*/*PP* combinations. Each isolate is virulent on one gene for resistance and not on the other.

5.

	I1	I2		*P1P1p2p2*	??*P2P2*
C1	LIT	LIT	*R1R1* ??	LIT1	LIT?
C2	HIT	LIT	*r1r1R2R2*	HIT	LIT2

Box 5 also allows the identification of two CGP's but not complete genotypes for the cultivars and isolates. The LIT? could result from either of the identified CGP's or from a third one.

6.

	I1	I2		*P1P1*??	?? ??
C1	LIT	LIT	*R1R1* ??	LIT1	LIT?
C2	LIT	LIT	?? ??	LIT?	LIT?

Box 6 provides little information. Anywhere from one to four CGPs could be involved.

7.

	I1	I2		??	??
C1	HIT	HIT	??	HIT	HIT
C2	HIT	HIT	??	HIT	HIT

Box 7 provides essentially no information. However, if a cultivar is known to have a gene for resistance, then both isolates carry the corresponding gene for virulence, and if an isolate is known to carry a gene for avirulence then the cultivar carries the corresponding gene for susceptibility.

Loegering et al. (1971) and Loegering and Burton (1974) developed a computer program that uses the seven boxes to analyze sets of data from the inoculation of groups of host genotypes with groups of pathogen isolates. In collecting the data, appropriate checks are used as standards in each test. For stem rust, Little Club was used as a standard check for a HIT. However, if the host lines were in some other genetic background, such as near-isogenic lines in Chinese Spring, then Chinese Spring was used. Any infection type clearly lower than that of the check in the same test was classified as low. If the check gave an IT4, an IT3 on another line would be low. Thus, the procedure detects genes having small effects. For the purpose of the computer program, only two infection types were recognized, high and low, coded as 0 and 1, respectively. However, on printouts the actual infection types are given for LITs. The program can do three things for a set of data: (1) group cultivars or isolates that have the same pattern of infection types, (2) compare cultivars or isolates of unknown genotype with those of known genotype, (3) postulate genotypes for the cultivars and isolates (Loegering 1984).

Browder and Eversmeyer (1980, 1982) developed a computer program to sort similar data toward a gene-for-gene model. To code infection types for leaf rust they used a system developed by Browder and Young (1975) involving two numbers and a letter for each IT. The two numbers represent the relative amount of sporulation and the relative pustule size on a 0 to 9 scale. The letters indicate such things as chlorosis or necrosis, but were not included in the sorting program. Once the data were sorted, they were examined visually, using the boxing procedure to identify corresponding gene pairs.

Roelfs et al. (1982a) used a computer system to handle a data bank produced by testing a large number of stem rust isolates on 42 "single gene resistances" in the host. When new rust cultures were tested, they were compared to previous cultures to see if they were similar or different. When a new host line was tested against 50 base cultures, the data were used to determine which genes the line cannot carry and which it may carry. If the line was more susceptible than a single gene line, it was assumed to not carry that gene. However, if the line gave the same infection type as a single gene line it could carry either that gene, or another gene, or a combination of genes that resulted in the same infection type. Thus, the system postulates the presence of more genes than are likely to be present in the unknown line. To determine those actually present, the unknown line must be crossed with lines carrying the postulated genes for resistance. However, the number of such crosses required is greatly reduced by the preliminary assessment.

Modawi et al. (1985) tested 20 near-isogenic lines carrying genes for leaf rust resistance and 30 winter wheat cultivars with 22 cultures of leaf rust. By comparing the data for the cultivars with those for the near-isogenic lines, genotypes were postulated for the cultivars. The method was useful in showing relationships between cultivars and in identifying possible new sources of resistance, but the hypotheses need to be tested by genetic studies.

Browder (1973) developed a slightly different method of postulating genotypes for rust resistance in wheat cultivars or lines, a method also based on the gene-for-gene hypothesis (he indicated that it was suggested by J.F. Schafer). Five wheat cultivars and 13 near-isogenic lines (NILs) carrying single genes for resistance to leaf rust were tested with 20 leaf rust cultures. The reactions of any two wheat genotypes to the 20 cultures can be compared and will fall into four classes. Both genotypes may give a low infection type (LIT:LIT), the first a LIT and the second a high infection type (LIT:HIT), the first a HIT and the second a LIT (HIT:LIT), or both a HIT (HIT:HIT). The presence or absence of each of the four classes will depend on the genes for resistance carried by the two genotypes and can be used to develop hypotheses about the genes they carry. However, as pointed out by Johnson et al. (1987), alternative hypotheses are possible.

Some of Browder's (1973) data are used as examples (Table 6.6). In the first example, every culture that is avirulent for *Lr10* is also avirulent for Tascosa and every culture that is virulent for *Lr10* is virulent for Tascosa. Thus, it can be hypothesized that Tascosa carries *Lr10*. As pointed out by Johnson et al. (1987), it

Table 6.6. The number of leaf rust cultures giving various combinations of low infection types (LIT) and high infection types (HIT) on pairs of wheat genotypes. (Data from Browder 1973)

	Number of cultures producing infection-type combinations			
Genotypes	LIT:LIT	LIT:HIT	HIT:LIT	HIT:HIT
Thatcher-Lr10:Tascosa	8	0	0	12
Thatcher-Lr10:Fox	8	0	9	3
Thatcher-Lr10:Blueboy II	8	0	12	0
Wichita-Lr1:Tascosa	5	7	3	5

is possible, although less likely, that Tascosa carries a different gene for resistance and by chance all cultures virulent for *Lr10* are also virulent for the gene in Tascosa, and all cultures that are avirulent for *Lr10* ar also avirulent for the gene in Tascosa. The two loci for virulence in the pathogen might be closely linked which would increase the possibility of such an occurrence.

In the second example, all cultures avirulent on *Lr10* were also avirulent on Fox. However, of those that were virulent for *Lr10*, nine were avirulent on Fox. Thus, Fox may carry *Lr10* plus a second gene for resistance. However, it is also possible that Fox carries only the second gene and the absence of cultures avirulent for *Lr10* and virulent for the gene in Fox is simply chance. The leaf rust cultures are probably not a random sample of the possible cultures.

The third example is similar to the second except that no combinations fell in the HIT:HIT class and none of the cultures was virulent on Blueboy II. This could be interpreted to mean that Blueboy II carries *Lr10* plus an additional gene or genes. If this is the case, the absence of cultures virulent for both genotypes was just chance – they should occur. However, as pointed out by Johnson et al. (1987), such a situation could also arise if Blueboy II does not carry *Lr10* but carries a gene or combination of genes that provides resistance to all 20 leaf rust cultures.

In the fourth example, all four combinations of infection types occur and it could be concluded that the two genotypes do not have a gene in common. Again as noted by Johnson et al. (1987), if Wichita-*Lr1* carried an unidentified second gene, then Tascosa could carry *Lr1* plus another gene and all four combinations would occur.

Additional evidence such as the similarity of infection types and the presence of genes in parents of genotypes being studied can be used to corroborate the hypotheses developed by the method (Browder 1973). Final proof of the presence of postulated genes may require appropriate testcrosses.

Rizvi and Statler (1982) tested 20 near-isogenic lines carrying single genes for leaf rust resistance and 20 hard red spring wheat cultivars with 20 leaf rust cultures. They used Browder's (1973) procedure to speculate about the genes carried by the cultivars and used infection types and pedigrees as additional evidence.

Statler (1984) did a similar analysis on a more extended set of wheat genotypes using 21 leaf rust cultures. He pointed out that a specific gene cannot be identified if the cultures are either all virulent or all avirulent on it. A wide range of virulence combinations is needed.

Johnson et al. (1987) discussed an example in which it had been erroneously concluded that 11 wheat cultivars carried *Yr8*. Kirmani et al. (1984) had tested the 11 cultivars with 40 cultures of yellow rust. Seven of the 40 were avirulent for *Yr8* and by chance all seven were also avirulent on the 11 cultivars. However, tests with other races and evidence from pedigrees indicated that the 11 did not carry *Yr8*.

Analysis of wheat cultivars and rust isolates can provide very useful information for the wheat breeder by grouping cultivars that probably carry the same genes for resistance, determining probable genotypes of potential parents and determining possible new sources of resistance.

6.4 Apparent Exceptions to the Gene-for-Gene Hypothesis

6.4.1 Introduction

Rust resistance is a complex character. Person and Sidhu (1971) surveyed 301 papers dealing with rusts (on all species, not just wheat). Resistance was reported to be due to dominant genes in 292 papers and recessive genes in 36. It was reported to be due to major genes in 291 papers and minor genes in 19. Gene interaction was reported in 43 papers and allelism in 27.

Since 1971, much more work has been done on more complex types of resistance such as slow rusting, and the number of papers reporting genes having small effects has increased greatly. For these types of resistance, because individual genes have small effects, it is much more difficult to identify them and to determine whether a gene-for-gene relationship exists. However, for barley leaf rust (*Hordeum vulgare: Puccinia hordei),* Parlevliet (1977) obtained evidence for differential interaction and suggested that minor genes operate on a gene-for-gene basis.

Apparent exceptions to the gene-for-gene hypothesis have frequently been reported. The exceptions can usually be resolved if the hypothesis is thought of in terms of the interaction of gene products rather than the interaction of genes.

6.4.2 Complementary Genes

Complementary genes for rust resistance have often been reported, but most examples are suspect. A number of these reports are based on field data with a mixture of races. A segregation suggesting complementary genes may be obtained because one gene gives resistance to some races and a second gene to others (Table 6.7). In reality the two genes are acting independently. In other cases, complementary genes have been hypothesized based solely on an F_2 segregation. However, confirmation is essential. In several cases, further testing has shown that the hypothesis was false. For example, several authors reported that cultivars Gabo, Lee, and Timstein had a pair of dominant, complementary genes for resistance to

Table 6.7. The origin of a 9:7 ratio in field tests with a mixture of races where one gene gives resistance to one or more races and a second gene gives resistance to one or more different races

Genotypes	Infection types with		
and frequency	Race 1	Race 2	Mixture
9 *R1-R2-*	LIT[a]	LIT	LIT
3 *R1-r2r2*	LIT	HIT	HIT
3 *r1r1R2-*	HIT	LIT	HIT
1 *r1r1r2r2*	HIT	HIT	HIT
Observed ratio	12 LIT:4 HIT (3:1)	12 LIT:4 HIT (3:1)	9 LIT:7 HIT

[a] LIT = low infection type and HIT = high infection type

stem rust. However, Loegering and Sears (1963) found that only one gene (*Sr11* on chromosome 6B) is involved, but it is transmitted normally through the eggs and only rarely through the pollen. They postulated that in Chinese Spring the *sr11* allele for susceptibility is closely linked to a gene, *Ki*, that causes the abortion of pollen carrying *ki*. Other cultivars do not carry *Ki* and a normal ratio of 3R:IS plant results. Luig (1968) suggested several other possible mechanisms for differential transmission of *Sr11*.

Nevertheless, dominant, complementary genes do occur. Baker (1966) demonstrated clearly the occurrence of dominant, complementary genes in oats (*Avena sativum*) for resistance to crown rust (*Puccinia coronata*). From the cross Bond/Fulghum he isolated susceptible lines which when intercrossed gave resistant plants. Similarly, in wheat Singh and McIntosh (1984a,b) identified dominant, complementary genes on chromosomes 4A ß and 3BS that conditioned resistance to leaf rust. In such cases it is probable that the two genes interact to produce a single active gene product that results in resistance.

6.4.3 Inhibitors

The occurrence of inhibitors of resistance has been reported for all three rusts, although in many cases results based on F_2 segregations have not been confirmed. In one of the clearest examples, Kerber and Green (1980) found that Tetra Canthatch (a tetraploid with the A and B genomes from Canthatch) and Canthatch nullisomic for chromosome 7D were resistant to some stem rust races to which normal Canthatch was susceptible. Evidently chromosome 7D carries a gene which inhibits the expression of a gene(s) on an A or B genome chromosome.

For leaf rust of wheat, inhibitors of genes for resistance have been reported in several cases. For example, McIntosh and Dyck (1975) found that a gene in Thatcher inhibited the resistance of *Lr23* to Canadian races but acted as only a partial inhibitor to Australian races. Kerber (1983) failed in repeated attempts to transfer leaf rust resistance from Stewart 63 durum wheat to Canthatch and Marquis. An amphidiploid between Stewart 63 and *Triticum tauschii* (D genome) was susceptible to leaf rust, indicating that the D genome carries a suppressor of resistance.

The occurrence of inhibitors can be accommodated by a slight modification of the gene-for-gene theory, again thinking in terms of gene products. The gene for resistance presumably produces an active gene product. The effect of the inhibitor must be to prevent the production of the gene product or to render it inactive. The pathogen should, therefore, have a gene corresponding to the resistance gene but not one corresponding to the inhibitor. However, this has not been tested.

6.4.4 Cumulative Effects of Genes for Rust Resistance

Earlier it was stated that when two or more effective genes for resistance are present, the gene giving the lowest infection type will usually be expressed. Actually, a number of cases have been reported where the combination of two or more genes results in a lower infection type than is conditioned by the most effective gene. In some cases the interaction involves identified genes but often the genes are unknown. For example, Schafer et al. (1963) intercrossed four cultivars having some resistance to leaf rust and selected stable lines that were more resistant than either parent. They then crossed ten of the resulting lines with both parents and showed that each line had genes from both parents. Similarly, Krupinsky and Sharp (1978) showed that minor genes for resistance to yellow rust could be combined to give high levels of resistance. Wallwork and Johnson (1984) intercrossed three cultivars that had incomplete resistance to yellow rust and obtained transgressive segregation. With stem rust the cultivar Thatcher gives a fleck reaction to race 56 of stem rust, although all of the genes it is known to carry condition higher infection types (Roelfs and McVey 1979). Thus, some of the genes must have cumulative effects.

In some cases, identified major genes have been shown to interact. Samborski and Dyck (1982) reported several cases in which pairs of genes for leaf rust resistance in a Thatcher background had cumulative effects. They emphasized the value of combining genes in wheat breeding. Loegering and Geis (1957) found that the genes *Sr6, Sr8,* and *Sr9* in Red Egyptian had cumulative effects, particularly at higher temperatures.

The fact that some genes act together to produce enhanced effects could mean that each pair of corresponding genes involves the interaction of different gene products, as would be expected. However, each interaction may have a different effect on the development of the pathogen and these effects could be cumulative.

6.4.5 Modifying Genes

Modifiers, that is genes that modify the effect of a resistance gene but have no direct effect on resistance themselves, have been reported frequently. In some cases the modifier may be an ineffective resistance gene. For example, Knott (1957) found that the resistance to race 15B-1 of stem rust conditioned by *Sr7a* is modified by genes *Sr9b, Sr10,* and *Sr11,* which are not effective against 15B-1. The gene *Sr2* in Hope and H-44 acts both as a modifier and as a gene with a cumulative effect (Knott 1968). With race 56, *Sr2* has no detectable effect in seedlings but conditions moderate resistance in adult plants. However, it increases the resistance conditioned by *Sr9d* (originally labeled *Sr1*) in both seedlings (a modifier effect) and adult plants (a cumulative effect). Although the effects of *Sr2* in combination with *Sr9d* would be labeled differently depending on whether it was observed in seedlings or adult plants, it is unlikely that there is any difference in the mechanism of the interaction.

Sometimes it appears that the genetic background rather than a specific modifier affects resistance. Thus, resistance may decline as a specific gene is transferred into a susceptible background by backcrossing. For example, Green et al. (1960) found that *Sr7a* and *Sr10* were less effective after five backcrosses to Marquis than in the donor parents. After four additional backcrosses some additional loss of resistance occurred with *Sr7a* (Knott and Green 1964). This suggests that either a number of genes are involved or one or more are linked to *Sr7*. Luig and Rajaram (1972) found that the resistance conditioned by *Sr5* changed from IT 0; to X to $X^{+}3$ after 5, 7, and 9 backcrosses to W2691. Knott (1981) found that the dominance or recessiveness of *Sr6* depended on genetic background, rust race, and environment (temperature and light). Under cool conditions *Sr6* was dominant with both race 56 and race 15B-1 in one cross, dominant with race 56 but recessive with race 15B-1 in a second, and recessive with both races in a third cross. The effect appeared to be due to the general genetic background rather than to a specific modifying gene.

Dyck and Samborski (1974) studied three alleles, *Lr2a*, *Lr2b*, and *Lr2c*, that condition resistance to leaf rust. When each allele was transferred into the susceptible cultivars, Thatcher, Red Bobs, and Prelude, by backcrossing, there were often small differences in infection type in the different genetic backgrounds. Each allele conditioned its lowest infection type in Thatcher, an intermediate type in Prelude and the highest in Red Bobs. When cultures from selfing race 11 of leaf rust were tested on the three backcross lines carrying one allele, in each case more cultures were classified as avirulent on the Thatcher backcross, fewer on the Red Bobs backcross and the least on the Prelude backcross. The segregation of virulent and avirulent cultures indicated that virulence on *Lr2a* was controlled by a single recessive gene, *p2*. However, virulence on *Lr2b* and *Lr2c* did not fit a single gene ratio. Dyck and Samborski (1974) postulated that on *Lr2b* and *Lr2c*, a second gene modified the expression of *p2*. However, the infection types for the cultures were essentially continuous which throws doubt on the classification.

6.5 Race-Specific and Race-Nonspecific Resistance

Vanderplank (1963) proposed that there are two different, distinct types of disease resistance in plants, vertical and horizontal. He defined vertical resistance as being effective against some races but ineffective against others. It is, therefore, race-specific, a term which I prefer. Thus, in race-specific resistance there is an interaction between genotypes of the host and genotypes of the pathogen. Vanderplank (1963) first defined horizontal resistance as being "evenly spread against all races of the pathogen". Thus, there is no genetic interaction between genotypes of the host and pathogen and the resistance is race-nonspecific. The terms vertical and horizontal arose from the figures used to illustrate the two types of resistance (Fig. 6.2). However, I prefer the terms specific and nonspecific, which are more descriptive in terms of the way the resistance functions.

The concept of vertical or specific resistance is clear. However, the concept of horizontal or nonspecific resistance is not so clear and has been subject to considerable controversy. Vanderplank has revised his definition over the years. In 1968 he stated that, "In horizontal resistance there is no differential interaction" (between varieties of the host and races of the pathogen) and that "an analysis of variance is one way of testing for differential interactions" (Vanderplank 1968). However, he also emphasized that a uniform ranking of host varieties in their resistance to different races of pathogen is the preferred method of detecting horizontal resistance. As an example, Vanderplank (1968) cited the data of Wellman and Blaisdell (1940) on the reaction of two tomato cultivars to five groups of isolates of *Fusarium oxysporum* f.sp. *lycopersici* (Table 6.8). The ranking of isolates is the same on both cultivars but clearly some isolates are more aggressive than others.

Vanderplank (1975) considered that "the distinction between differential and nondifferential interactions is clear and unambiguous", i.e., the distinction between specific and nonspecific resistance. This is true only if all possible types of the

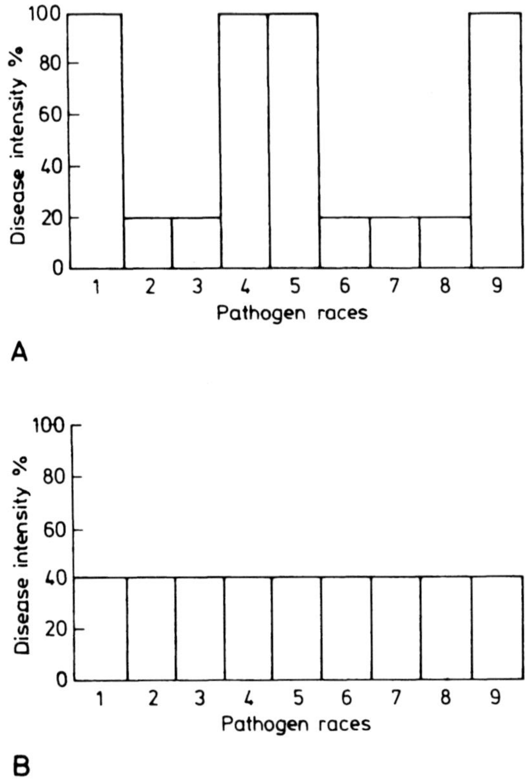

Fig. 6.2. Diagrams to illustrate vertical (**A**) and horizontal (**B**) resistance as originally described by Vanderplank (1963)

Table 6.8. The reaction of two varieties of tomato to five groups of isolates of *Fusarium oxysporum* f.sp. *lycopersici*. (Vanderplank 1968)

Isolate group[a]	Reaction severity[b] on Bonny Best	Marglobe
Fully raised (fluffy)	10.4	7.5
Raised sclerotial	8.7	6.3
Intermediate raised	8.3	4.5
Intermediate appressed	6.3	4.1
Appressed (slimy)	4.7	3.0

[a] Based on their type of growth in culture.
[b] On a 0 to 15 scale with 0 indicating no disease and 15 severe disease.

pathogen have been tested so that there is no possibility of ever discovering a new pathogen genotype that results in a differential interaction between host and pathogen. Resistance can only be classified as nonspecific as long as no pathogen genotype is discovered that can overcome it. An additional problem may occur if the differential interactions are small and are obscured by large uncontrolled variability.

6.5.1 Theoretical Models for Specific and Nonspecific Resistance

Parlevliet and Zadoks (1977) attempted to develop a concept of disease resistance that integrated horizontal and vertical resistance. In their Table 1, they present three models involving four cultivars and four pathogen isolates (Table 6.9). In the first model (A), the four cultivars maintain a uniform ranking with the four isolates. All variation is due to the main effects of cultivars and isolates, there is no interaction sum of squares, and resistance is horizontal. The cultivars differ in their levels of resistance and the isolates differ in aggressiveness.

In the second model (B), each cultivar and each isolate shows the same four disease severities in a fixed pattern that results in all eight means being the same.

Table 6.9. Parlevliet and Zadok's (1977) models showing the disease severities for all combinations of four host cultivars and four pathogen isolates

	Isolates						
Cultivars	1	2	3	4	Means		
A. Main effects only – horizontal resistance							
I	10	20	30	40	25	Sums of squares	
II	20	30	40	50	35		
III	30	40	50	60	45	Total S.S.	4000
IV	40	50	60	70	55	Cultivar S.S.	2000
						Isolate S.S.	2000
Mean	25	35	45	55	40	Interaction S.S.	0
B. Interaction effects only – vertical resistance							
I	10	30	50	70	40	Sums of squares	
II	30	50	70	10	40		
III	50	70	10	30	40	Total S.S.	8000
IV	70	10	30	50	40	Cultivar S.S.	0
						Isolate S.S.	0
Mean	40	40	40	40	40	Interaction S.S.	8000
C. Main effects and an interaction – a combination of horizontal and vertical resistance							
I	10	30	40	20	25	Sums of squares	
II	40	20	30	70	40		
III	30	30	50	50	40	Total S.S.	4400
IV	40	60	60	60	55	Cultivar S.S.	1800
						Isolate S.S.	1000
Mean	30	35	45	50	40	Interaction S.S.	1600.

Thus, the main effects due to either cultivars or isolates are 0 and all the variation is due to the interaction, i.e., resistance is vertical.

The third model (C) combines the first two – there is variation among the cultivars and among the isolates, and there is an interaction. Thus, it combines horizontal and vertical resistance.

Parlevliet and Zadok's (1977) model for vertical resistance (B) is unrealistic. It is unlikely that any combination of genetically variable cultivars and isolates will result in similar mean disease severities for either all the cultivars or all the isolates. Experience with near-isogenic wheat lines carrying single genes for stem rust resistance indicates that adult plants of a line will tend to give the same disease severity in the field with all rust isolates to which it conditions resistance, assuming that disease intensity is similar in each test (Knott unpublished). A model which is based on actual data and should be realistic is given in Table 6.10. Even though the resistance is entirely due to genes for specific resistance, there is variation due to both host genotypes and races, in addition to the interaction. Thus the model appears to combine both horizontal and vertical resistance, although in fact the resistance is entirely due to genes for vertical or specific resistance. Any combination of diverse host and pathogen genotypes is almost certain to result in significant main effects for both, even if all of the genes involved are specific. While the presence of an interaction may indicate the occurrence of specific resistance, the presence of significant main effects does not necessarily indicate the occurrence of horizontal resistance.

Table 6.10. A model for stem rust severities for all combinations of four near-isogenic wheat lines and four rust races, based on actual data from field tests with single races

	Races						
Host genotypes	56	15B-1	29-1	11	Mean	Sums of squares	
Marquis	80	80	80	80	80.0	Total S.S.	13,393.75
Marquis-Sr6	10	10	80	10	27.5	Cultivar S.S.	5618.75
Marquis-Sr7a	80	40	80	40	60.0	Races S.S.	768.75
Marquis-Sr9a	30	80	30	80	55.0	Interaction S.S.	7006.25
Mean	50.0	52.5	67.5	52.5	55.6		

Robinson (1980) developed a model for horizontal resistance in which the effects of the genes in the host and pathogen were multiplicative, although constant ranking was maintained. Robinson's model, with every second host genotype and pathogen genotype eliminated to simplify it, is given in Table 6.11. The model produces some surprising results. Even though it has no genes for resistance, the most susceptible host type is immune to one pathogen type and shows a range in disease severity from 0 to 100%. The most resistant host type is immune to all pathogen types. The most aggressive pathogen type is avirulent on one host type and produces a range in disease severity from 0 to 100%. The least aggressive pathogen type is avirulent on all host types. The sums of squares calculated from the model show sizable effects due to pathogen types and host types, and also a sizable interaction.

Table 6.11. A model for horizontal resistance. (After Robinson 1980). The figures in the body of the table are disease severities

Pathogen types – percent of (+) alleles[a]2	Host types – percent of (-) alleles[a]2 100	80	60	40	20	0	Means
100	100	80	60	40	20	0	50
80	80	64	48	32	16	0	40
60	60	48	36	24	12	0	30
40	40	32	24	16	8	0	20
20	20	16	12	8	4	0	10
0	0	0	0	0	0	0	0
Means	50	40	30	20	10	0	25

Sums of squares	
Total S.S.	25,900
Pathogen types S.S.	10,500
Host types S.S.	10,500
Interaction S.S.	4900

[a]The effects of the (-) alleles in the host and the (+) alleles in the pathogen are multiplicative.

If the model is realistic (and the postulated multiplicative effects may not be) then data showing a similar pattern should be obtained from field tests with single, pure isolates. In this case the conclusion, based on an analysis of variance, would be that both horizontal and vertical resistance were present. However, constant ranking would indicate that only horizontal resistance was operating.

Parlevliet and Zadoks (1977) also set up models for polygenic resistance, an addition model, and an interaction model. In the addition model, each host gene for resistance decreases disease severity by 10% and each pathogen gene for virulence increases disease severity by 10%. The genes in the host and the pathogen act independently of each other. For example, if the host carries four genes for resistance, reducing disease severity by 40%, and the pathogen carries two genes for virulence, increasing disease severity by 20%, the net result is a decrease of 20%. In the interaction model, host genes for resistance act as in the addition model. However, pathogen genes for virulence have an effect only if the host carries the corresponding gene for resistance, i.e., they act in a gene-for-gene manner. The results for the two models, but reduced to three loci in each of the host and pathogen for simplicity, are given in Table 6.12. One problem arises because a host with no resistance genes is assumed to show 100% disease severity. All pathogen genotypes will produce 100% disease severity on a host with no genes for resistance, regardless of the number of genes for virulence they carry, i.e., additional genes for virulence have no effect. Because of this, an analysis of variance on the addition model will give a significant interaction mean square, even though there is no change in ranking with different host genotypes. A comparison of the analyses of variance for the two models shows that there is little difference (Table 6.13). The interaction mean square is somewhat larger for the interaction model. The main difference between the two is that there are no changes in rank in the addition model but there are some small changes in the interaction model. There is no way of knowing how realistic the models are or whether both systems may operate. Small interactions may be difficult to detect because of experimental error. However, they were detected by Parlevliet (1977) in the barley:barley leaf rust system, although the interaction effects were small compared to the total variation in resistance.

Table 6.12. Addition and interaction models for polygenic host: pathogen systems. (After Parlevliet and Zadoks 1977). Genes for resistance are shown as dominant and genes for virulence as recessive. Heterozygous genotypes are not considered

A. Addition Model

	Pathogen genotypes								
Host genotypes	*PaPa PbPb PcPc*	*papa PbPb PcPc*	*PaPa pbpb PcPc*	*PaPa PbPb pcpc*	*papa pbpb PcPc*	*papa PbPb pcpc*	*PaPa pbpb pcpc*	*papa pbpb pcpc*	Means
r1r1r2r2r3r3	100	100	100	100	100	100	100	100	100.0
R1R1r2r2r3r3	80	100	100	100	100	100	100	100	97.5
r1r1R2R2r3r3	80	100	100	100	100	100	100	100	97.5
r1r1r2r2R3R3	80	100	100	100	100	100	100	100	97.5
R1R1R2R2r3r3	60	80	80	80	100	100	100	100	87.5
R1R1r2r2R3R3	60	80	80	80	100	100	100	100	87.5
r1r1R2R2R3R3	60	80	80	80	100	100	100	100	87.5
R1R1R2R2R3R3	40	60	60	60	80	80	80	100	70.0
Means	70.0	87.5	87.5	87.5	97.5	97.5	97.5	97.5	100

B. Interaction Model

	Pathogen genotypes								
Wheat genotypes	*P1P1 P2P1 P3P3*	*p1p1 P2P2 P3P3*	*P1P1 p2p2 P3P3*	*P1P1 P2P2 p3p3*	*p1p1 p2p2 P3P3*	*p1p1 P2P2 p3p3*	*P1P1 p2p2 p3p3*	*p1p1 p2p2 p3p3*	Means
r1r1r2r2r3r3	100	100	100	100	100	100	100	100	100.0
R1R1r2r2r3r3	80	100	80	80	100	100	80	100	90.0
r1r1R2R2r3r3	80	80	100	80	100	80	100	100	90.0
r1r1r2r2R3R3	80	80	80	100	80	100	100	100	90.0
R1R1R2R2r3r3	60	80	80	60	100	80	80	100	80.0
R1R1r2r2R3R3	60	80	60	80	80	100	80	100	80.0
r1r1R2R2R3R3	60	60	80	80	80	80	100	100	80.0
R1R1R2R2R3R3	40	60	60	60	80	80	80	100	70.0
Means	70.0	80.0	80.0	80.0	90.0	90.0	90.0	90.0	100

Table 6.13. Analysis of variance for the addition and interaction models in Table 6.12

Source	Degrees of freedom	*Addition model* SS	MS	*Interaction model* SS	MS
Total	63	13575	215.5	14 400	225.0
Host genotypes	7	5475	782.1	4800	685.7
Pathogen genotypes	7	5475	782.1	4800	685.7
Interaction	49	2625	53.6	4800	98.0

As pointed out by Parlevliet and Zadoks (1977), in the addition model all mutations to virulence are expressed. In the interaction model, only those mutations for virulence that correspond to a gene for resistance in the host are expressed. As a result, the average disease severity is lower for the interaction model than for the addition model. Thus, an interaction model is more favourable for the host and the addition model for the pathogen.

Person et al. (1983) criticized Parlevliet and Zadok's (1977) polygenic models and their conclusions because the addition model had a theoretical maximum of 200% disease severity even though the maximum possible is 100%. The theoretical maximum for the interaction model was 100%. They developed their own models based on the assumption that polygenes in both host and pathogen acted equally and additively and, therefore, generated "potential" variability in each organism that was continuous and fitted a normal distribution. The potential variability in each organism cannot be measured separately but is measurable only in the interaction between the host and pathogen. Person et al. (1983) postulated both additive and multiplicative models for the interaction between host and pathogen to obtain realized levels of disease reaction. Both models produced constant ranking of the host genotypes for all pathogen genotypes. The additive model resulted in a normal distribution of disease reactions whereas the multiplicative model resulted in a skewed distribution with a peak on the resistant side. For the additive model, a shift in the distribution for one organism produced a much smaller shift in the disease reactions, i.e., there was a "phenotypic damping". Natural selection will tend to favor increased virulence in the pathogen and increased resistance in the host. These changes will move the disease reactions in opposite directions and will tend to balance each other.

A problem with all of these models is the difficulty of determining whether the assumptions made are realistic or not. They also illustrate how difficult it is to distinguish between specific (vertical) and nonspecific (horizontal) resistance.

6.6 Durable Resistance

The difficulties with the concept of horizontal or nonspecific resistance led Johnson (1979, 1981, 1983) to propose the term "durable resistance". He defined durable resistance as, "resistance that has remained effective in a cultivar during its widespread cultivation for a long sequence of generations or period of time, in an environment favorable to a disease or pest". The concept has the advantage that it describes what has actually been observed but does not imply any underlying cause or genetic basis. It focuses on the fact that is of great practical importance, the durability of the resistance and, therefore, its value in breeding.

The basis for durable resistance has been the subject of considerable debate (see for example papers in Lamberti et al. 1983). The simplistic answer is that there are two possible reasons for resistance being durable.

1. Either the pathogen is unable to develop a virulent race or virulent races are not competitive with the prevalent avirulent races.

2. For some reason virulent races do not come into contact with the resistant host.

Why either of these possibilities occurs is much more complex.

The interaction between wheat and a wheat rust is complex and the durability of a resistance may be affected by many factors.

1. Genetic. Very rarely it appears that a rust pathogen has great difficulty in producing a mutant for virulence on a particular gene. As described earlier, the gene *Sr26* derived from *Agropyron elongatum* Host Beauv. has remained effective against stem rust in Australia since its release in a cultivar in 1971 (Luig 1983).

In other cases, particular combinations of genes have remained effective for long periods. In Australia the cultivars Timgalen, Timson, Cook, and Shortim contain several genes for stem rust resistance and were not overcome by changes in the rust population for many years (Luig 1983). Their resistance may have been due to the combination of *Sr6* and *SrTt1* (*Sr36*). Although neither gene alone conferred durable resistance, no isolate with the combined virulence appeared for a long period. The reason for the failure of this combination to appear quickly is not known, but it may have been because the use of resistant cultivars kept the rust population small, thus reducing the chance of mutants becoming established.

In Canada, Selkirk, which carries at least six genes for stem rust resistance, has remained resistant since its release in 1953, although it is no longer grown to any extent.

In some cases, but not always, resistance that is of the quantitative type and controlled by several genes with small, cumulative effects has proven to be durable. For example, the resistance to yellow rust of Cappelle-Desprez has remained effective in England while that of Maris Bilbo and Hobbit has not (Johnson 1983). All three have quantitative resistance and the latter two were originally more resistant than Cappelle-Desprez. However, races appeared that could attack Maris Bilbo and Hobbit, although not Cappelle-Desprez. The results indicated that the quantitative resistance of Maris Bilbo and Hobbit is race-specific and probably controlled by single genes. The resistance of Cappelle-Desprez is more complex (Johnson and Lupton 1987).

In Kenya, the cultivars Kenya Page and Africa Mayo remained resistant to stem rust in the field although they were susceptible to several prevalent races in the seedling stage (De Pauw 1978). Each has at least two genes for post-seedling resistance, including *Sr2* (Luig 1983).

2. The life cycle of the rust. A sexual cycle is not known for yellow rust and is not important in the life cycle of leaf and stem rust in some areas of the world such as Australia and the main wheat areas of North America. In these areas new gene combinations must be produced by either mutation or possibly somatic recombination. If a cultivar carries two or more genes for resistance for which the rust population in the area lacks the corresponding genes for virulence, two or more simultaneous mutations will have to occur in the rust to produce virulence. The probability of such a mutant type occurring and becoming established may be small. Alternatively, the virulent rust genotype will have to develop in two or more steps. Whether this can happen and how long it takes will depend on the virulence of the intermediate genotypes on the cultivars being grown in the area.

3. Gene deployment. In North America, stem and leaf rust of wheat overwinter in the south, particularly in Texas and Mexico. In the spring and summer they travel north over 3000 km along the "Puccinia Path". If cultivars with different resistance genes are grown in different sections of the Puccinia Path, the rusts are subject to continually changing selection pressure. The stem rust surveys in the United States and Canada show that the frequency of virulence on specific *Sr* genes changes as the rust moves from south to north. For example, in 1975, 38% of the isolates from area 1 (the overwintering area) were virulent on *Sr9a* but only 4% in areas 6 and 7 (the northern great plains area of the United States) (Roelfs and McVey 1976). On the other hand, only 19% of the isolates in area 1 were virulent on *Sr11* compared to 91% in areas 6 and 7. In the adjacent area of Canada, 6.9% were virulent on *Sr9a* and 89.5% on *Sr11*. In 1976, 27% of the isolates from the southern states (areas 1 and 2 were virulent on *Sr6* compared to only 2% in the northern states (areas 6 and 7) (Roelfs et al. 1977). In the same year, virulence on *Sr9a*, *Sr9b*, *Sr15* and *Sr17* also decreased during the passage north, while virulence on *Sr9e*, *Sr10*, *Sr11*, *SrTt1* (*Sr36*) and *SrTmp* increased. The wheat cultivars grown in the Puccinia Path show considerable genetic diversity with large areas planted to soft and hard red winter wheats, soft white wheats, red spring wheats and durum wheats. Although the genes that they carry for rust resistance are often not known, there is undoubtedly considerable diversity. This is probably the reason for the changes in frequencies of rust races from area to area. If a resistance gene is not used in the southern U.S.A., virulence on it will not build up and it should be safe to use the gene in Canada. Even if a virulent rust genotype develops in Canada, the spores will still have to be blown back into the overwintering area in the southern U.S.A., and this may not happen.

4. Epidemiology. If weather conditions are unfavorable for the pathogen, fewer infections will develop on the host and the spore load will be reduced. This may be particularly important if the host cultivar has only a moderate level of resistance.

6.7 Association and Dissociation of Genes for Virulence

Vanderplank (1975) noted the apparent absence of combined virulence on *Sr6* and *Sr9d* in stem rust in Canada, although the combination was common in the United States and Mexico. He concluded that virulence genes and combinations of virulence genes can affect fitness of rust genotypes. In 1978 he concluded that certain virulence genes repel each other and are, therefore, dissociated, e.g., virulence for *Sr6* and *Sr9d* (Vanderplank 1978). Since virulence for *Sr6*, *Sr9a*, and *Sr9b* mutually repel virulence for *Sr9d* and *Sr9e*, they will tend to be associated. However, Green (1981) showed that the apparent absence of combined virulence for *Sr6* and *Sr9d* was not true but resulted from the use in Canada of a tester that carried both *Sr9d* and a second, unknown gene, *H*. Thus, rust isolates that were virulent for *Sr6* and appeared to be avirulent for *Sr9d*, were really virulent for *Sr9d* but avirulent for *H*. The dissociation between virulence for *Sr6* and *Sr9d* in Canada did not occur.

Vanderplank (1982, 1984) analyzed North America stem rust data and concluded that genes for virulence could be divided into two groups. Genes within a group associate while genes in opposite groups dissociate. For each group there is a matching group of resistance genes in the host, the ABC and XYZ groups. To determine the groups, Vanderplank calculated the expected frequency for each pair of virulence genes based on the frequencies for the individual genes. For example, the observed frequencies for *p6* and *p7b* in 1975 were 0.11 and 0.82, giving an expected frequency for the combinations of 0.11 x 0.82=0.09. The observed frequency was 0.02, indicating that *p6* and *p7b* dissociate and *Sr6* and *Sr7b* are therefore, in opposite groups.

The method used by Vanderplank (1982, 1984) is based on the assumption that North America stem rust is a ramdomly mating population. Wolfe and Knott (1982) noted some of the problems that can arise when survey data which do not meet this assumption are analyzed. Knott (1984) and Roelfs and Martens (1984) pointed out that the stem rust population in the Puccinia Path of North America is largely asexually reproduced. In fact, one race, 15-TNM, has dominated the population and in 1975 made up 68% of the rust isolates. Race 15-TNM has the virulence formula (effective/ineffective genes)*Sr6, 9a, 9b, 13, 15, 17/5, 7b, 8a 9c, 9d, 10, 11, Tt1(36), Tmp*. It is avirulent for each of the ABC genes (*Sr6, 9a, 9b, 15* and *17*) and virulent for each of the XYZ genes (*Sr7b, 9e, 10, 11, Tt1, Tmp*). Thus, the apparent association and dissociation of virulence genes in the North American rust population is due to the prevalence of one rust race.

Vanderplank (1985) has attempted to rebut the arguments of Knott (1984) and Roelfs and Martens (1984). He argues that the same pattern holds in Mexico, where race 15-TNM is unimportant. While it is true that the pattern holds for the two examples he gives, it does not hold for many other pairs of genes for virulence. He also argues that the pattern holds for both asexual and sexual populations in North America. In fact, the results of Alexander et al. (1984) show that one sexual cycle reduces the strength of the associations and dissociations exactly as would be predicted by population genetics theory. All of Vanderplank's observations can be simply explained by examining the structure of the stem rust populations involved. Typically they are made up of a limited number of asexually propagated genotypes with little opportunity for recombination. Furthermore, the methods of collecting rust samples, such as collecting from farmers' fields, may result in a nonrandom sample of the population. Any rust genotypes that are not virulent on the cultivars being grown will be discriminated against. When samples are taken from different locations over a large area, it is unlikely that they represent a single, homogeneous population. Thus, the use of population genetics methods to analyze rust survey data is very questionable (Wolfe and Knott 1982).

CHAPTER 7

Cytogenetic Analysis of Resistance

7.1 Development of Aneuploids

Because bread wheat is a polyploid, many types of aneuploids (plants that do not have the normal chromosome number or some multiple of it) are viable and fertile. Sears (1954) produced the complete sets of 21 monosomics ($20'' + 1'$, lacking one member of a chromosome pair), 21 nullisomics ($20''$, lacking both members of one pair), 21 trisomics ($20'' + 1'''$, carrying three chromosomes of one type) and 21 tetrasomics ($20'' + 1''''$, carrying four chromosomes of one type).

All of the original aneuploids produced by Sears (1954) arose in the cultivar Chinese Spring, primarily as the progeny of either haploid plants or nullisomic 3B plants which are partially asynaptic. These plants gave rise mainly to monosomics, which produced nullisomics on selfing, and trisomics, which produced tetrasomics. During meiosis monosomic chromosomes not infrequently misdivide at the centromere and give rise to telocentric chromosomes which have only one arm and occasionally isochromosomes which have two identical arms.

All 42 possible telocentric chromosomes have been obtained, 41 in Chinese Spring and telo 7DL in Canthatch (Sears and Sears 1979). Twenty-five monotelosomics ($20'' + t'$, where t = a telosome) are maintained, including at least one for each chromosome, except 2A and 6B (Sears and Sears 1979). Thirty-four ditelosomics ($20'' + t''$), including at least one for each chromosome, are fertile, although their fertility varies greatly. Various lines carrying both telocentrics derived from one chromosome are available for all chromosomes except 7D, i.e., dimonotelosomics ($20'' + t'' + t'$), double ditelosomics ($20'' + 2t''$) and double monotelosomics ($20'' + 2t'$) (Sears and Sears 1979).

The various aneuploids, particularly monosomics and telocentrics, have been used extensively to identify the chromosomes carrying particular genes in wheat and to map them relative to the centromere. The wealth of aneuploids has permitted some very elegant cytogenetic studies to be carried out.

7.2 Maintaining Monosomics and Other Aneuploids

When a monosomic plant undergoes meiosis, it might be expected that half the gametes would carry the monosomic chromosome and half would lack it. However, because the monosomic chromosome does not have a homolog with which to pair, it often fails to move normally to a pole during one or the other of the two meiotic divisions. As a result, about half the time the monosomic chromosome is not included in a nucleus and appears as a micronucleus in the pollen tetrad. Therefore, only about 25% of the gametes carry all 21 chromosomes and about 75% carry only 20 chromosomes. Furthermore, when a monosomic plant is selfed, the 20-chromosome pollen frequently fails to function, the frequency of functioning varying from about 1 to 19% depending on the particular chromosome (Morris and Sears 1967). The range is due to the fact that some chromosomes are more essential to the growth of the pollen than others. The result is that about 73% of the progeny of a monosomic plant are monosomics (Table 7.1). Thus, monosomic lines are easily maintained by selfing. Since monosomic plants tend to show some sterility, their spikes should be bagged to prevent any accidental outcrossing.

In reproducing monosomics, there is always the possibility of a univalent shift occurring. In a line that is monosomic for one chromosome, a plant may appear that is monosomic for a different chromosome. This can occur when a normal chromosome pair fails to pair at meiosis, something that happens occasionally. One of the unpaired chromosomes may not get into a nucleus and a 20-chromosome gamete that lacks a chromosome other than the one that was originally monosomic will result. When it is fertilized by a 21-chromosome gamete, a plant monosomic for a different chromosome than the original line will have been produced. Monosomic stocks must be checked regularly by crosses to the corresponding ditelosomic lines. If the monosomic is correct, the monosomic F_1 progeny will have a configuration of 20″ + t′. If univalent shift has occurred the monosomic progeny will be 20″ + 1t″ + 1′, i.e., the telosome will be paired with a normal chromosome and the univalent will be a whole chromosome. As a precaution, in all monosomic studies a minimum of two monosomic plants per line should be used. It is often wise to use plants from two monosomic lines for each chromosome. It is unlikely that more than one plant in a line will have undergone monosomic shift.

With the development of monotelosomic lines, the problem of monosomic shift can be prevented for most chromosomes. In a monotelosomic line, a monosomic shift will result in a plant that is 20″ + 1′ rather than 20″ + t′. The plant can be identified and discarded. Most monotelosomics are fairly similar to the monosomic for the same chromosome and can be maintained and used in crosses. However, Chinese Spring monotelosomics for chromosomes 4A, 6A, and 2D give low seed set and are difficult to use (R.S. Kota personal communication). The monotelosomics for the

Table 7.1. The gametic types in monosomic wheat plants, their frequency of functioning, and the progeny from self-pollinating a monosomic plant. (After Morris and Sears 1967)

		Pollen	
Eggs	Frequency (range)	21 chromosomes 96% (81–99)	20 chromosomes 4% (1–19)
21 chromosomes	25% (14–29)	21″ – Disomic 24% (11–29)	20″ + 1′ – Monosomic 1% (0.1–5)
20 chromosomes	75% (61–86)	20″ + 1′ – Monosomic 72% (49–85)	20″ – Nullisomic 3% (0.6–16)

In total – 24% (11–29) disomic, 73% (49.1–90) monosomic and 3% (0.6–16) nullisomic.

group 3 chromosomes are early and those for 4D and 5D are late so that some adjustment in seeding dates may be necessary.

Nullisomics occur with varying frequencies in the progeny of monosomics. They can often be recognized by their lack of vigor and narrow leaves. Unfortunately, most nullisomics are almost completely male sterile. In Chinese Spring, nullisomics 1A, 1D, 3A, 3B, 3D, 6A, 6B, and 7D are the most fertile and can be maintained and used in crosses (Law et al., 1987). Their chromosome configurations must be checked frequently, since unexpected types such as compensating trisomics are not infrequent. A plant may be nullisomic for one chromosome but trisomic for a homoeologous chromosome. Because of their low fertility, bagging to ensure selfing is essential.

Trisomic plants are reasonably fertile and give about 45% trisomics and 3% tetrasomics (Sears 1954). Tetrasomics are relatively stable, with most producing more than 80% tetrasomic progeny. However, a few produce only about 50% tetrasomics and a correspondingly high frequency of trisomics (Sears 1954).

Plants carrying telocentrics or isochromosomes are fairly frequent in the progeny of monosomics. Monotelosomic and monoisosomic plants are more vigorous than the corresponding nullisomics but still vary greatly in vigor and fertility depending on the chromosome and the arm involved.

7.3 Producing Monosomic Series in Other Cultivars

Once the monosomic series had been produced in Chinese Spring, it could be used to produce a set of monosomics in any other cultivar. For many purposes a genetic background other than Chinese Spring is desirable. The procedure involves crossing each of the 21 Chinese Spring monosomics as females with the cultivar. Monosomic plants are selected and backcrossed as females to the cultivar and the process repeated. To ensure that most of the genotype of the recurrent parent is recovered, a minimum of five backcrosses is necessary. Only 1/64 (1.6%) of the genes should then be from Chinese Spring. To test for the presence of such genes, Law et al. (1987) suggest selfing monosomic plants to produce disomic lines and comparing the lines with the recurrent parent. Alternatively, each monosomic can be produced in duplicate, starting with the first cross. After a number of backcrosses, the duplicate lines for each chromosome can be compared. Both members of a pair will carry the same monosomic chromosome and any differences will be due to genes from the donor parent carried on other chromosomes.

A danger in producing a new monosomic series is that univalent shift may occur during the backcrossing, i.e., an initial cross is made with one monosomic line but one or more of the progeny are monosomic for a different chromosome. For example, in a plant monosomic for 1A, a chromosome pair, e.g., 2B, may occasionally either fail to pair at meiosis or may not separate normally. As a result, a gamete carrying 1A but lacking 2B can be produced and give rise to a monosomic 2B plant. Pairing failures are more common in crosses than in a cultivar. The identity of the final monosomic plants should be checked by crossing them with the corresponding ditelosomics. Producing duplicate monosomic lines for each chromosome can be used as a form of insurance. It is unlikely that both will undergo univalent shift.

Another problem may arise if the cultivar differs from Chinese Spring by a reciprocal translocation. In two of the 21 crosses, the monosomic F_1 plants will carry one normal chromosome and two chromosomes with translocations, and at meiosis most cells will have a trivalent. Depending on chromosome disjunction from the trivalent, the trivalent may persist during the backcrossing and it may be difficult to obtain plants that are homozygous for one translocation and monosomic for the other.

Full sets of monosomic lines are now available in many cultivars (listed by McIntosh 1987b).

7.4 Using Monosomics to Identify Chromosomes Carrying Genes for Rust Resistance

Monosomic analysis is most commonly used when the inheritance of rust resistance in a line or cultivar has already been determined. If the inheritance of the resistance is not known, it may prove to be too complex for analysis. Monosomic analysis works best when only one or two genes are involved.

The first step in a monosomic analysis is to use a set of 21 monosomic lines in a susceptible cultivar as the female parents in crosses to the rust-resistant parent. The monosomic lines must be used as females in order to obtain a high frequency of monosomic F_1 plants, since 20-chromosome pollen functions with a low frequency. In order to have a control, a cross should also be made to the susceptible disomic parent. Although the Chinese Spring monosomics are often used, monosomics in some other cultivar may be more suitable. For example, if the resistant parent is winter habit, it may be simpler to use a monosomic series in a winter cultivar and vernalize all of the parents at the same time. Since the monosomic parents do not breed true, it is necessary to determine the chromosome numbers of the plants in segregating progenies. This can be done by either root-tip counts at mitosis or pollen mother cell counts at meiosis. The latter is somewhat safer, since pairing can be observed to make sure that some other type of aberration has not occurred, but both are used.

The F_1 progeny of the 21 crosses are grown out, tested for their rust reaction, and chromosome counts made. The F_1 plants should be grown under very good growing conditions, since monosomic plants tend to be sterile. For the same reason it is also very important to bag the heads to prevent outcrossing. If resistance is due to a single recessive gene (*r*) which expresses itself when hemizygous (*r-* in which the dash represents a missing chromosome), it should be possible to identify the chromosome in the F_1 generation. In the 20 crosses in which the monosomic chromosome does not carry the locus involved, all of the progeny, both monosomic and disomic, will be heterozygous *Rr* and, therefore, susceptible. These are referred to as the noncritical crosses. In the one remaining cross, the critical cross, the disomic plants will be *Rr* and susceptible, but the monosomic plants will be *r-* and resistant, since the gene is effective when hemizygous. Only the critical cross will segregate in the F_1 generation.

Unfortunately, this is not a common situation because recessive genes are often ineffective when hemizygous, and identification of the chromosome involved usually requires considerably more work (see below).

If no cross segregates in the F_1 generation, F_2 families are grown from monosomic F_1 plants of each of the 21 crosses. If a single dominant gene is involved, then in the 20 noncritical crosses a normal ratio of 3 resistant: 1 susceptible plant will be obtained. In the critical cross the F_1 plants will be *R*-. When selfed they will give F_2 plants that are *RR* (disomic), *R*- (monosomic) and -- (nullisomic). Only the nullisomics will be susceptible except that occasionally the univalent will misdivide, the arm carrying *R* will be lost and a susceptible plant that is either 20″ + t′ or 20″ + i′ (i = an isosome) will result. The exact ratio of resistant and susceptible plants that is obtained will depend largely on the frequency with which the 20-chromosome pollen functions. If it functions about 4% of the time, then the ratio will be 97R:3S or about 32:1 (Table 7.2). Usually the ratio is a little lower, perhaps 15:1, either because the 20-chromosome pollen functions with a frequency higher than 4% or because misdivision of the univalent results in some additional susceptible plants. In any case, there will be an excess of resistant plants. The susceptible plants will often be nullisomics which are usually weak and narrow-leaved and, therefore, distinguishable. When a preliminary indication of the critical cross has been obtained, more F_2 progeny from the cross can be grown. It is also useful to do meiotic chromosome counts on susceptible plants to confirm that they are either 20″ or a product of misdivision of the univalent such as 20″ + t′ or 20″ + i′. Monotelosomic or monoisosomic plants can be used to determine the chromosome arm carrying the gene for resistance. The plants are crossed to the known Chinese Spring ditelocentrics to determine which arm is missing. The missing arm must carry the gene for resistance.

An example of an F_2 monosomic analysis which showed that *Sr7a* is on chromosome 4B is given in Table 7.3. Not infrequently, even when it is known that only one gene is involved in resistance, two or even three monosomic crosses will give probabilities of less than 0.05 of fitting the expected 3:1 ratio, as in Table 7.3. This, of course, is not surprising. When 20 noncritical crosses are tested, there is a sizeable probability that one or more will give a probability below 0.05 just by chance. The chance of obtaining a noncritical line with a probability below 0.05 can be greatly reduced by growing F_2 progeny from at least two monosomic F_1 plants. It is unlikely

Table 7.2. Results expected in a cross between the critical monosomic line and a parent carrying a single dominant gene for rust resistance (R). The frequencies will vary somewhat depending on the chromosome involved and the frequency of functioning of 20-chromosome pollen

	Male gametes	
Female gametes	21 chromosomes - 96% *R*	20 chromosomes - 4% -
21 chromosomes - 25% *R*	21″ - disomic *RR* 24%	20″ + 1′ - monosomic *R*- 1%
20 chromosomes - 75% -	20″ + 1′ - monosomic *R*- 72%	20″ - nullisomic -- 3%

Ratio 97% (*RR* or *R*-): 3% (--)

Table 7.3. The segregation for reaction to race 15B of stem rust in F_2 seedlings from crosses between the Chinese Spring monosomic lines and Kenya Farmer which carries *Sr7a*. (Knott 1959)

Monosomic chromosome	Number of F_2 plants Resistant	Susceptible	P for a 3:1 ratio
1A	77	35	0.10–0.20
2A	51	7	0.02–0.05*
3A	29	10	1.0
4A	27	12	0.30–0.50
5A	58	20	1.0
6A	62	16	0.30–0.50
7A	40	15	0.50–0.95
1B	11	6	0.30–0.50
2B	25	12	0.20–0.30
3B	53	22	0.30–0.50
4B	160	0	< 0.001**
5B	77	20	0.30–0.50
6B	61	19	0.50–0.95
7B	30	9	0.50–0.95
1D	13	6	0.50–0.95
2D	22	13	0.05–0.10
3D	29	9	1.0
4D	47	11	0.20–0.30
5D	13	7	0.30–0.50
6D	64	16	0.30–0.50
7D	25	14	0.10–0.20

*and ** The two crosses differ significantly from the expected 3:1 ratio at the 0.05 and 0.001 levels of probability, respectively.

that both will give segregations with probabilities below 0.05. If a single F_2 progeny is grown for each cross, then the probabilities of obtaining 0, 1, 2, . . . crosses out of 20 that give probabilities below 0.05 by chance can be obtained by expanding the binomial, $(p+q)^{20}$, where $p=0.95$ and $q=0.05$. The expected probabilities are:

0 crosses – 0.358
1 cross – 0.377
2 crosses – 0.189
3 crosses – 0.060
4 or more crosses – 0.016.

Just by chance, about 38% of the time one of a set of 20 noncritical monosomic crosses will give a probability of fitting a 3:1 that is below 0.05, and about 19% of the time two crosses will. The deviation may occur in either direction but if a cross has an excess of susceptible plants, the deviation will almost certainly be due to chance. If two crosses show an excess of resistant plants, one will often have a much larger excess than the other and is probably the critical cross. Usually confirmation can be obtained by growing more plants of the two crosses. The chances are that a second sample of one cross will fit a 3:1 ratio but the critical cross will continue to have an excess of resistant plants. Further confirmation can be obtained by doing meiotic chromosome counts on the susceptible plants. In the critical cross they will be nullisomic or carry a product of misdivision of the univalent, while in the noncritical cross they can be disomic, monosomic, or even nullisomic.

A monosomic analysis may be more complicated if the gene being studied is only partially dominant and conditions an intermediate infection type, such as IT2. In this case, a continuous range of infection types from IT2 to IT4 will probably occur in the crosses, with no clear separation into classes. It is not valid to arbitrarily divide the plants into classes by calling ITs 2 and X as resistant and 3 and 4 as susceptible, and attempting to fit the segregations to a ratio. The classes are unlikely to coincide with genotypes. However, if a cross to the susceptible disomic has been made, the results from it can be used as a standard with which to compare the results from 21 monosomic crosses. The appropriate procedure is not a chi-square goodness-of-fit test, as has sometimes been used in the literature, but rather a chi-square test for association (a contingency chi-square). This is because the standard is not a theoretical ratio but an actual set of data which is itself subject to error. The appropriate test is to see whether the two segregations could have both come from the same population. To illustrate, suppose that the results in the cross to the susceptible cultivar were 42R:28S plants and in the cross to the monosomic IA line were 38R:17S. The resulting chi-square from a contingency table would be 2.275 and would be non-significant, i.e., both segregations could be from the same population (Table 7.4). If a goodness-of-fit test had been done using the results from the cross to the susceptible disomic as a theoretical ratio, the chi-square would have been 4.103 and the deviation would be significant (P = 0.01–0.05). The results for each monosomic cross are tested in the same way.

Table 7.4. A chi-square test for association (a contingency chi-square) comparing the F_2 segregations in crosses of a resistant parent with a monosomic line and with the susceptible disomic

	Number of plants		
Parent	Resistant	Susceptible	Totals
Disomic	42 (46.15)[a]	28 (23.85)	70
Monosomic	47 (42.85)	18 (22.15)	65
Totals	89	46	135

[a] The expected frequencies are in brackets.
Chi-square = 2.275, P = 0.10–0.25.

If the gene for resistance is recessive and not effective when hemizygous, a further modification of the procedure is necessary. In this case neither the F_1 nor F_2 results will identify the critical cross. In the F_2 progeny of the critical cross, the segregation will be approximately 1R:3S as in the noncritical crosses (Table 7.5). The 24% of the plants that are disomic will be *rr* and resistant, the 73% that are monosomic will be *r*- and susceptible, and the 3% that are nullisomics will be -- and susceptible. The ratio will be about 24R:76S, or close to 1:3, and not distinguishable from the noncritical crosses. However, the critical cross can be identified by counting the

Table 7.5. The segregation in the F_2 progeny of monosomic F_1 plants in the critical cross involving a recessive gene for rust resistance

	Male gametes	
Female gametes	96% *r*	4% -
75% -	72% *r*-(S)	3% -- (S)
25% *r*	24% *rr* (R)	1% *r*-(S)

Ratio 24% R: 76% S

chromosomes in resistant F_2 plants. In the critical cross, all the resistant plants will be disomic, whereas in the noncritical crosses about 73% of the resistant plants will be monosomic. If necessary, the rust reactions and chromosome counts can be confirmed by growing out small F_3 families from resistant F_2 plants.

If two genes control resistance, then monosomic analysis becomes more difficult and the procedure depends on the type of gene action as follows:

1. If two independent, dominant genes are involved, both giving resistance to the race being used, the expected F_2 ratio in the 19 noncritical crosses is 15R:1S. In the two critical crosses, the ratio will be > 15R:1S, but will usually be indistinguishable from a 15:1. It is sometimes possible to determine the critical crosses by doing chromosome counts on the susceptible F_2 plants. In the 19 noncritical crosses they can be disomics, monosomics, or nullisomics. In the two critical crosses they can be only nullisomics or misdivision products such as monotelosomics. Alternatively, if monosomic F_1 plants are used as pollen parents in crosses to a susceptible parent, the 19 noncritical lines will give ratios of 3R:1S, whereas the two critical lines will give mostly resistant plants. Sometimes if the resistant parent is tested with a number of rust cultures, cultures can be found that are virulent for one gene but not for the other. Such cultures can be used to locate the genes since only one gene will segregate.

If the two genes are completely dominant and condition distinctly different infection types such as IT0 and IT2, the segregation within F_2 families may help in identifying one of the critical crosses. The 19 noncritical crosses will segregate 12 IT0:3 IT2:1 IT4. In one critical cross almost all of the plants should be IT0 and only the occasional plant will be IT2 or IT4. Thus, it should be identifiable. However, the second critical cross should give a segregation similar to the noncritical crosses.

2. If two dominant, complementary genes condition rust resistance, the expected ratio in the 19 non-critical crosses is 9R:7S. In the two critical crosses the expected ratio is about 3R:1S. These two ratios can be distinguished if F_2 families of at least 100 plants are tested. As a final check, F_3 families can be grown from resistant F_2 plants in the apparently critical crosses. In a critical cross, about 2/3 of the families should segregate about 3R:1S and 1/3 should be all or mostly resistant. No families will segregate 9R:7S. In a noncritical cross, about 4/9 of the F_3 families from resistant F_2 plants should segregate 9R;7S, 4/9 3R:1S, and 1/9 all R.

3. If one dominant and one recessive gene condition rust resistance, the expected ratio in the 19 noncritical crosses is 13R:3S. In the critical cross for the chromosome carrying the dominant gene, most of the plants should be resistant.

However, in the critical cross involving the chromosome carrying the recessive gene, the ratio will be about 13R:3S and the cross will not be detectable. If chromosome counts are done on the susceptible F_2 plants in the latter cross, they will be either monosomic or nullisomic. In the noncritical crosses, about 24% of the susceptible plants will be disomic.

If the two gene are either linked or on the same chromosome but more than 50 crossover units apart, then only one monosomic cross will be critical. No crossing over can occur in F_1 plants monosomic for the critical chromosome. The critical cross will behave as though only one gene is segregating and most plants will be resistant. Whether the critical chromosome can be identified from F_2 segregations will depend on the segregation in the noncritical crosses, and that will depend on the type of gene action involved and the recombination percentage.

These examples cover most of the possibilities that may arise in monosomic studies. If the inheritance of resistance is more complicated, the individual genes should be separated into separate lines before monosomic analyses are done.

7.5 Chromosome Mapping Using Telocentrics

Once the chromosome carrying a gene (*R*) for rust resistance has been identified, the chromosome arm carrying the gene and the gene's distance from the centromere can be determined using telocentrics (Sears 1962, 1966a). Assuming that both ditelocentric lines are available for the chromosome carrying the gene for rust resistance, both are crossed with the homozygous resistant line carrying the gene whose location is to be mapped. The F_1 plants are then crossed to a susceptible male parent with normal chromosomes. In the cross with the line ditelocentric for the chromosome arm not carrying the gene, no crossovers between the gene and the centromere will occur (Fig. 7.1A). All resistant plants will be 21″ and all susceptible plants 20″ + t1″. The gene is, therefore, on the opposite chromosome arm. In the cross with the second ditelocentric line, crossing over will occur between the gene and the centromere and the frequency will depend on the distance they are apart (Fig. 7.1B). If, for example, the two noncrossover types had frequencies of 42 and 38, and the crossover types had frequencies of 11 and 9, the crossover frequency would be $20/100 = 0.20$ and *R* is 20 map units from the centromere. However, if the gene is close to the centromere, no crossing over may occur between them. Sears (1972b) noted that in telocentric chromosomes crossing over is reduced near the centromere which can bias map distances.

If only one ditelocentric line is available for the chromosome carrying a resistance gene, then it may not be possible to obtain a complete chromosome mapping. If the testcross to a susceptible parent produces no apparent crossovers, the resistance is probably on the opposite arm. It is possible, however, that it is on the arm being tested but so close to the centromere that no crossing over occurred. If crossovers do occur, then normal mapping can be done.

The methods outlined above require that chromosome counts be done on all plants and this is time-consuming. One way of reducing the amount of cytology is to make use of the fact that transmission of a telocentric through the pollen is often low

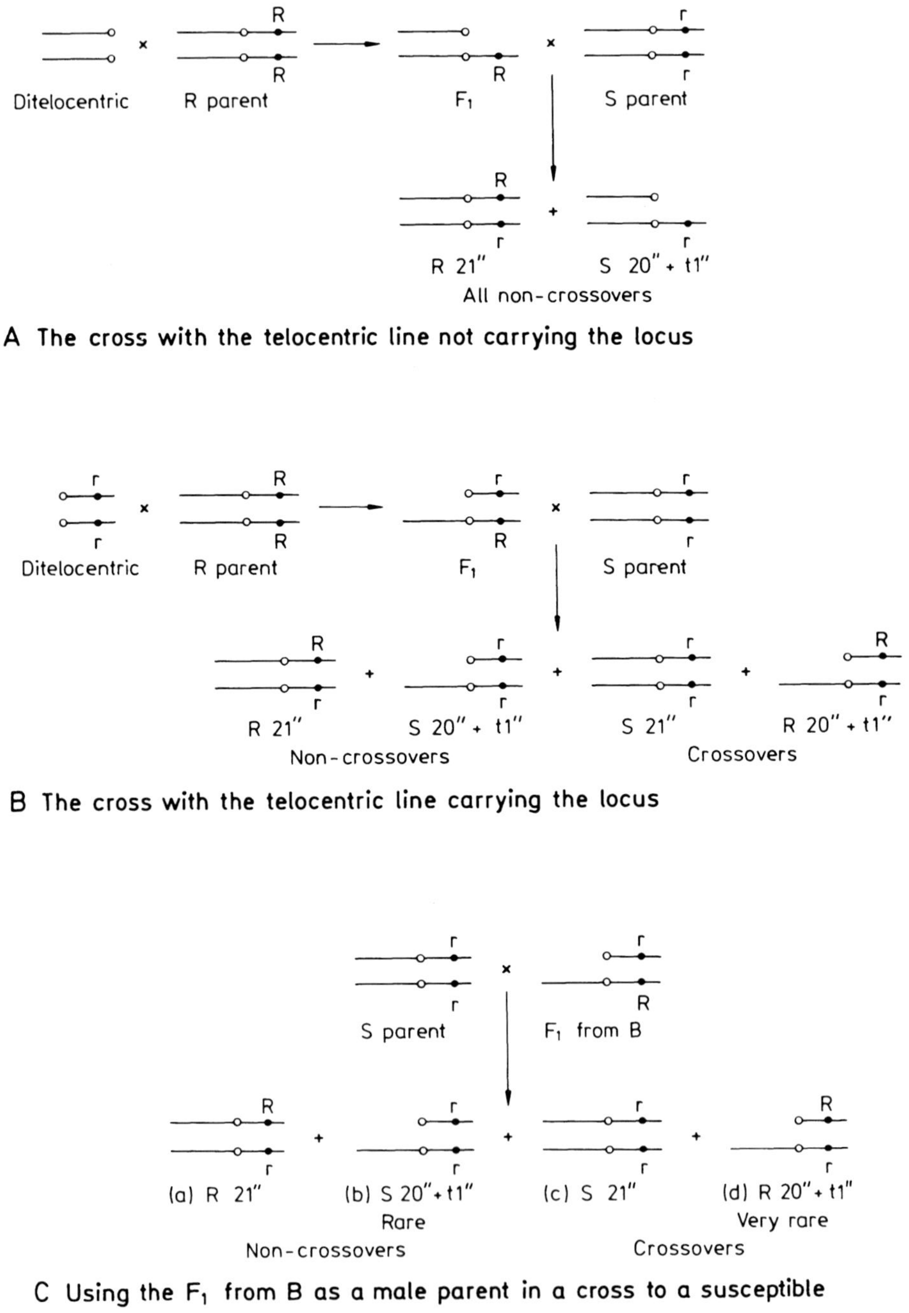

Fig. 7.1. Using ditelocentric lines to map the location of a gene for resistance relative to the centromere (see the text for description of the procedure)

(Sears 1966). The F_1 from the cross with the ditelocentric carrying the locus is used as the male parent in the cross to a susceptible (Fig. 7.1C). Most of the progeny will receive the full chromosome from the F_1 plant and will be resistant when rust-tested. Only the susceptible plants are studied cytologically. Most will be 21″ and will have resulted from a crossover, while a few will be 20″ + t1″ as a result of having received the unchanged telocentric. The plants that are susceptible and 21″ constitute the major crossover class. The second crossover class will be resistant and 20″ + t1″, but will have a low frequency because of the low transmission of the telocentric through the pollen. Since no cytology is done on the resistant plants, this class will be included with the resistant plants that are 21″. An approximate estimate of crossing over can be made by dividing the frequency of the major crossover class by the sum of its frequency plus the total frequency of resistant plants. For example, suppose that the crossover frequency is 20% and telocentric transmission is 10%. In the F_1 plants, the four types of pollen will be produced in frequencies of 40, 40, 10, and 10 but will function to give the four types of progeny (a, b, c, d in Fig. 7.1C) in frequencies of 40, 4, 10, and 1. The crossover estimate will be 10/(10 + 40 + 1) = 19.6%, only slightly less than the real value of 20%. Sears (1966) gives a number of examples of using telocentrics to map genes.

7.6 Production and Use of Substitution Lines

Once a monosomic series has been produced in a cultivar A, it is possible to substitute each of the chromosomes of cultivar B for its homolog in cultivar A (Sears 1953, Unrau et al. 1956, Kuspira 1966). The 21 monosomic lines of cultivar A are used as female parents in crosses with cultivar B. Monosomic F_1 plants from the 21 crosses are used as pollen parents in backcrosses to the corresponding monosomic lines of cultivar A (Fig. 7.2). Since 20-chromosome pollen rarely functions, the monosomic progeny of the backcross should be monosomic for a chromosome from cultivar B and lacking the homologous chromosome from cultivar A. The backcrossing can be continued for any number of generations. The number of backcrosses is important. With a limited number of backcrosses some genetic material from the donor, in addition to the chromosome being transferred, will be retained. The monosomic progeny can then be selfed to obtain disomic plants. These plants will carry 20 pairs of chromosomes of cultivar A plus one pair from cultivar B substituted for the corresponding pair of cultivar A. Since the chromosome from cultivar B has always been monosomic, it has had no chance to cross over with its homolog from B. In this way a set of 21 substitution lines can be produced.

Two types of problems can arise in the above method of developing substitution lines. First, univalent shift can occur in the monosomic lines of cultivar A, so they should be checked regularly by crosses to the corresponding ditelosomic lines. Even then, monosomic shift may have just occurred in the plant used in a cross or even in the egg that is fertilized to give rise to a hybrid seed. After the final backcross has been made, the monosomic plants should be crossed with the corresponding ditelosomic to check that a univalent shift has not occurred. Secondly, occasionally during the backcrossing a 20-chromosome male gamete will function

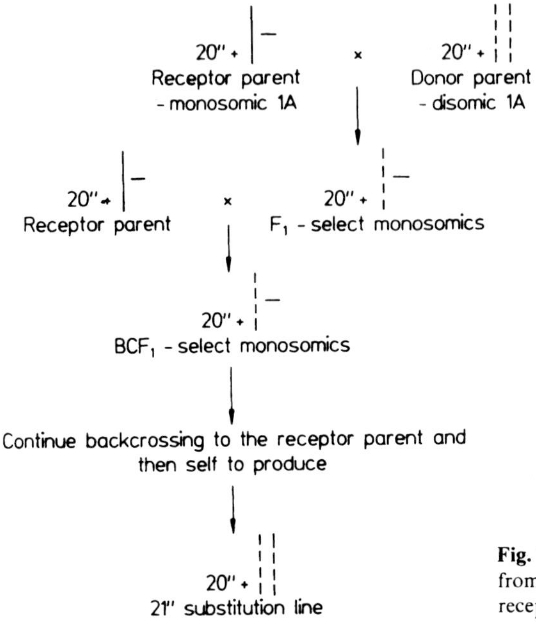

Fig. 7.2. The substitution of chromosome 1A from a donor parent for chromosome 1A of a receptor parent

and fertilize a 21-chromosome egg. This results in a monosomic plant that is monosomic for the chromosome of cultivar A instead of the chromosome of cultivar B, sometimes called a monosomic switch. Unfortunately, this is difficult to detect unless the chromosome of cultivar B carries a known marker gene, the loss of which can be observed.

There are several ways of avoiding these problems. If any nullisomics of cultivar A are fertile enough to be reproduced, they can be used in place of the corresponding monosomics. In Chinese Spring, nullisomics 1B, 7B, and 7D are the most fertile. If nullisomics are used, the hybrids will always be monosomic for the correct chromosome. If another chromosome is lost, the resulting hybrid will be $19'' + 2'$ and can be discarded. If any chromosome of the donor parent carries an easily recognized marker gene, it can be used to determine the presence of that chromosome. Unfortunately, good markers are not common. If monotelosomics, monoisosomics, or even double monotelosomics are available in the receptor parent, they can be used instead of monosomic lines. Then if a 20-chromosome male gamete functions and produces a monosomic plant it will be $20'' + t'$ instead of $20'' + 1'$ and can be identified and discarded. If univalent shift occurs in the receptor parent, the monosomic hybrids will be $19'' + t1''$ and can also be identified and discarded.

If the donor and receptor parents differ by a translocation, complications can arise which have been described by Kuspira (1966).

The production of substitution lines is too time-consuming to be justified just to determine the location of genes for rust resistance in a cultivar. They are more likely to be produced to study more complex, quantitative characters. However, if a rust-resistant cultivar has been used as a donor to produce substitution lines in a susceptible cultivar, the lines can be used to study the inheritance of rust resistance.

Such studies will be most successful in identifying independent genes that have sizeable effects. For example, Sears et al. (1957) studied four sets of substitution lines in which chromosomes from Hope, Thatcher, Red Egyptian, and Timstein had been substituted into Chinese Spring. In tests using a total of 26 cultures of stem rust, nine chromosomes were shown to carry genes for resistance.

If resistance is complex, tests of substitution lines may not be of value. For example, if two genes are complementary, then neither substitution will have any effect on rust resistance. If resistance is polygenic, it is possible that no substitution will have a detectable effect unless several genes are located on one chromosome.

Knott (1986a) studied the Chinese Spring substitution lines carrying chromosomes from Hope and Thatcher, for reaction to stem rust race 56 in the field. Although Hope and Thatcher have good resistance to race 56, none of the substitution lines showed a similar degree of resistance. Hope had resistance genes on chromosomes 3B (*Sr2*) and 4D, and Thatcher possibly on 6A and 3B (*Sr12*). However, individual genes apparently had small effects and several had to be combined to give the resistance of the parents.

When substitution lines are used in the study of quantitative characters, the data often reveal how complex their inheritance is. For example, Kuspira and Unrau (1957) studied Chinese Spring lines with chromosomes substituted from Thatcher, Hope, and Timstein. Five Thatcher substitutions had a higher percent protein than Chinese Spring (the Hope and Timstein substitutions were not tested). Ten Thatcher substitutions, nine Hope substitutions, and 13 Timstein substitutions headed earlier than Chinese Spring. However, this does not mean that five chromosomes carry genes that specifically control protein content or 9–13 chromosomes carry genes that specifically control heading date. More probably each chromosome carries genes that affect the development of the plant in various ways so that protein content is indirectly affected in the first case or maturity in the second.

Tests on substitution lines will indicate which chromosomes carry genes controlling a particular character but not the inheritance of the character. Unrau (1958) proposed a method for studying the inheritance once a particular substitution line had been identified. The substitution line is crossed with the cultivar in which the substitution was made. The F_1 plants should be homozygous for all genes except those on the substituted chromosome from the donor and its homolog from the recipient cultivar. These two chromosomes will pair and undergo crossing over, resulting in the production of recombinant chromosomes. These recombinant chromosomes are then isolated in individual lines as follows: If the substituted chromosome is, for example, 1A, then the F_1 plants are used as males in a testcross to monosomic 1A or monotelosomic 1A of the recipient cultivar. Monosomic plants are selected. The univalent chromosomes in the selected plants will have resulted from pairing and crossing over between the substituted chromosome and its homolog in the F_1 plants. If a set of these testcross plants is selfed, the disomic progeny of any one plant will be homozygous for a specific recombinant chromosome and for the other 20 chromosomes of the recipient cultivar. In this way, a series of lines, each derived from a different testcross plant and homozygous for a different recombinant chromosome, can be produced. If the substituted chromosome carries three genes affecting the character under study, then there are $2^3 = 8$ possible types. The homozygous lines derived from the testcross plants are studied to analyze the recombinant chromosomes and determine the number of genes involved. The method can be used to study the inheritance of several characters, not just one, and to determine linkages among the genes involved. For example, Law and Wolfe (1966) analyzed the substitution of Hope 7B in Chinese Spring and showed that it carried a gene for mildew resistance closely linked to a gene having a small effect on ear emergence time.

The method is particularly useful for quantitative characters. Law (1967), and Law and Johnson (1967) analyzed the same Chinese Spring (Hope 7B) lines and identified single genes on Hope 7B that affected height, grain weight, grain number and tiller number, resistance to leaf rust and culm colour. Resistance to leaf rust was shown to be influenced also by genes on the homoeologous chromosomes 7A and 7D. The method was also used by Worland and Law (1986) in the identification of gene *Yr16* for resistance to yellow rust.

7.7 The Use of Reciprocal Monosomic Hybrids

Chromosome substitution had been suggested as a possible method of wheat breeding. However, because of the length of the procedure, a method was needed to determine the possible contribution of a chromosome before it was considered for substitution. McEwan and Kaltsikes (1970) proposed the use of reciprocal monosomic crosses to provide the information. For example, using Chinese Spring and Rescue monosomics, for each chromosome two crosses are made, e.g., Chinese Spring (CS) monosomic 1A × Rescue (Rsc), and Rescue monosomic 1A × Chinese Spring. Monosomic F_1 plants from the two crosses have the constitutions 20′CS/20′Rsc + 1′(1A Rsc) and 20′CS/20′Rsc + 1′(1A CS), respectively. Thus the two types are identical except for the 1A chromosomes they carry. Differences between them are due to the different potencies of the two 1A chromosomes for the character under study. McEwan and Kaltsikes (1970) demonstrated differences in yield and days to heading caused by chromosomes from several parents.

The method has limited applicability because it requires that a monosomic series be available in both parents. However, it has been used to identify the chromosomes responsible for differences between cultivars in several quantitative characters.

7.8 The Backcross Reciprocal Monosomic Method

One of the limitations of monosomic analysis is that it does not work well with quantitative characters. In F_2 families from monosomic F_1 plants, there will be continuous variation. It is then difficult to distinguish between the effects of the segregation of alleles in noncritical crosses and the effects of chromosome dosage (monosomics and disomics) in critical crosses.

Snape and Law (1980) developed a method using backcross reciprocal monosomic lines to study quantitative characters. The method requires a monosomic series in only one cultivar and may be useful for studying rust resistance that is complex in inheritance. The first step is to cross a set of 21 monosomic lines in one cultivar (A) with a second cultivar (B) carrying the character to be analyzed. Monosomic F_1 plants, which will be monosomic for a chromosome from cultivar B, are selected and backcrossed reciprocally to the original monosomic parent (Fig. 7.3). When an F_1 plant is used as the male, almost all of the functional gametes will carry 21 chromosomes. Therefore, the monosomic plants from the backcross will be monosomic for the chromosome from B. In the reciprocal backcross, however, the monosomic chromosome will be from cultivar A. Thus, two sets of backcross F_1 families are produced that have the same (although segregating) genetic background for 20 chromosomes, but differ for the monosomic chromosomes from the two parents. The two types of backcross F_1 families can be compared to determine the effects of the chromosomes from the two parents. If sufficient numbers of each type are compared, the effects of the segregating genetic background will average out. The backcross F_1 families will include about 25% of disomic (21″) plants which will tend to dilute the differences between pairs of families. To avoid this problem, monosomic backcross F_1 plants can be selfed and disomic F_2 plants selected for comparison. All of the genes on the chromosomes being compared will be homozygous, thus eliminating any possible complication from hemizygosity and the effects of monosomy.

Snape et al. (1983) used the backcross reciprocal monosomic method to study the effects of chromosomes carrying genes for semi-dwarfism in the cultivar Sava. Previously work had implicated chromosomes of homoeologous groups 1, 2, and 5, without locating a major dwarfing gene. The appropriate crosses were made for the chromosomes of these three groups, except for 2B. For each backcross, four disomic F_2 plants were extracted and their families grown in the field. The results indicated that chromosome 2D carried a major dwarfing gene and 1B a minor gene.

The method was also applied to the cultivar Avalon, using Chinese Spring monosomics, to identify a gene on chromosome 3B, probably derived from Hybrid

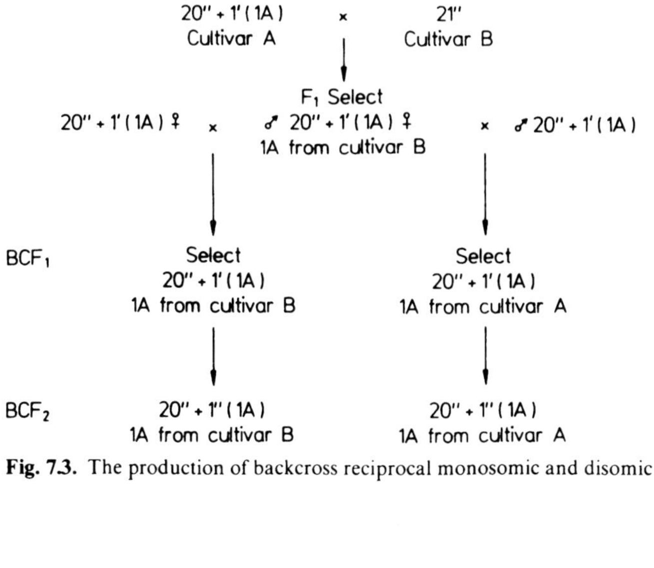

Fig. 7.3. The production of backcross reciprocal monosomic and disomic lines

46, that gave resistance to race 41E136 but not race 37E132 of yellow rust (Worland and Law 1987).

As yet the procedure does not appear to have been used to study the inheritance of complex disease resistance.

7.9 Monosomic Analysis in Durum Wheat

Joppa (1987) reviewed the use of aneuploid analysis in tetraploid wheat.

In theory it should be possible to use the A and B genome monosomic lines in a bread wheat to locate genes in durum wheat. The crosses are usually easily made and the hybrids reasonably fertile. The A and B genome chromosomes should pair normally and genes on them segregate in normal ratios. However, the 34-chromosome plants (13″A or B + 1′A or B + 7′D) often show poor fertility and segregations may be disturbed in the F_2 populations. Attempts by the author (unpublished) to locate genes for stem rust resistance in several durum lines were not successful. However, several authors have identified chromosomes carrying genes, particularly for quantitative characters, by comparing monopentaploid (34 chromosome) with pentaploid (35 chromosome) plants (reviewed by Joppa 1987). For the method to be successful, genes must have a different effect when hemizygous than when homozygous.

Several workers have attempted to produce a set of monosomic lines in durum wheat. Mochizuki (1968) crossed the A and B chromosome monosomics of Chinese Spring with Stewart durum and backcrossed to Stewart. He obtained 12 of the possible 14 monosomics. However, the vigor and fertility of monosomic plants was low. Even when the monosomic plants were pollinated with normal durum pollen, the frequency of monosomics averaged only 11.14% (range 2.6–30.2%). Hanchinal and Good (1982) found even lower transmission rates in a different genetic background.

Because of the problems with monosomic lines, there has been interest in other aneuploids that might be useful in genetic analysis, such as substitution–monosomics in which one D genome chromosome has been substituted for its A or B genome homoeolog, e.g., 13″ + 1′(1A) + 1′(1D). Joppa and Williams (1977) crossed Chinese Spring lines nullisomic for an A or B genome chromosome and tetrasomic for the compensating D genome chromosome, with the durum wheat Langdon. Five backcrosses were made to Langdon and all 14 compensating substitution-monosomics were developed. When backcrossed to Langdon, the substitution-monosomics produced an average of 33% of substitution-monosomics. They were more vigorous and fertile than monosomics. However, when selfed they produced 0–31% (mean of 7.1%) substitution-monosomics (Salazar and Joppa 1981). Thus, it is best to maintain the lines by crossing to a normal durum.

The lines can be used to identify chromosomes carrying particular genes in much the same way that monosomics are used in bread wheat (Joppa and Williams 1977). Substitution-monosomic plants of each of the 14 lines are crossed to the cultivar to be analyzed. If resistance is due to a single, dominant gene, the F_2 populations from substitution-monosomic F_1 plants in the 13 noncritical crosses should segregate 3R:1S. In the critical cross, most of the F_2 plants should be resistant since pollen lacking the critical chromosome functions rarely. However, in a few cases such as the 4D-4A monosomic substitution, the D genome chromosome compensates better for the missing chromosome in the pollen and more susceptible plants will occur. In this case, detecting the critical cross may be difficult (Joppa 1987). The monosomic D genome chromosomes will have no effect unless one of them carries a gene, such as an inhibitor, which affects resistance. The D genome chromosomes are present to compensate for the missing A or B genome chromosomes and improve vigor and fertility.

Salazar and Joppa (1981) used the substitution-monosomics in a slightly different way to locate genes for stem rust resistance in Langdon. Since the lines were produced in Langdon, they were crossed to the susceptible cultivar, Marruecos-9623. If resistance in Langdon is due to a dominant gene, in the 13 noncritical crosses F_1 plants that are $13'' + 2'$ should produce F_2 families that segregate 3R:1S. In the critical cross, the F_2 progeny should be all susceptible, since the F_1 plants that are $13'' + 2'$ do not carry the critical chromosome from Langdon. If two genes are involved, the ratio should change from 15R:1S to 3R:1S and if three genes are involved from 63R:1S to 15R:1S. Despite some complications, Salazar and Joppa (1981) concluded that Langdon has a gene for resistance to stem rust culture 0r9e on chromosome 7A, three genes for resistance to 111-SS2 on 2B, 3B and 7A, and two genes for resistance to GB121, one on 4A and one probably on 6A.

More recently Joppa and Williams (1988) developed D-genome disomic substitutions from the substitution monosomics. The degree of compensation of the D-genome chromosome for the homoeologous A or B genome chromosome varied considerably. Some lines were reasonably fertile, others were sterile, and the 4D(4B) substitution did not survive. However, the sterile or inviable lines could be maintained if a monosome or telosome of the A or B genome chromosome was present. Konzak and Joppa (1988) used the disomic substitutions to identify the chromosome (7B) carrying a recessive gene for chocolate chaff.

CHAPTER 8

Breeding Methods

8.1 Introduction

In breeding for rust resistance, the objective of the wheat breeder is to develop cultivars that will remain resistant at least for the period when they are grown commercially. The achievement of this goal is affected by many factors involving the host (e.g., the type of resistance being used), the pathogen (e.g., the life cycle of the rust in the area), the environment (e.g., the favorableness of the weather for rust development), and man (e.g., the decision as to which cultivars will be grown). Many breeding systems have been and are being used in different wheat-growing areas depending on the specific situation.

The tremendous economic value of growing rust-resistant cultivars has been demonstrated many times. Each time that resistance breaks down in an area and an epidemic occurs the importance of breeding for resistance is demonstrated anew.

8.2 Field Rust Nurseries

Because of the large amount of material to be handled in a breeding program, rust testing is usually done in field nurseries. They have the advantage that the plants are tested under natural field conditions. In a few areas natural epidemics can be depended on, but in most cases artificial inoculation is required. Good conditions should be provided for plant growth and the spread of the rust. This may require some irrigation. However, if the normal commercial crop is rainfed, selection under irrigated conditions for characters other than rust resistance may not be desirable.

Most nurseries include rust susceptible spreaders about every 10 to 15 rows (Fig. 8.1). They are used for inoculation and to insure that there is ample spore production and spread of the rust. The more frequent the spreader rows are, the more uniform the epidemic should be. The spreader rows can be planted with a single susceptible cultivar but it is often useful to mix two or three cultivars of different maturities in order to spread the epidemic over a longer period. Commercial cultivars of known rust reaction are commonly planted at regular intervals to provide a check both for rust reaction and other characters. Segregating generations such as the F_2 are usually planted in long rows with the plants spaced 15 to 30 cm apart. In a pedigree system, early generation, single plant progenies are grown in short rows of 30 to 50 plants. In later generations when most lines are reasonably homozygous, fairly thickly seeded 1 or 2 m rows or even hill plots are adequate.

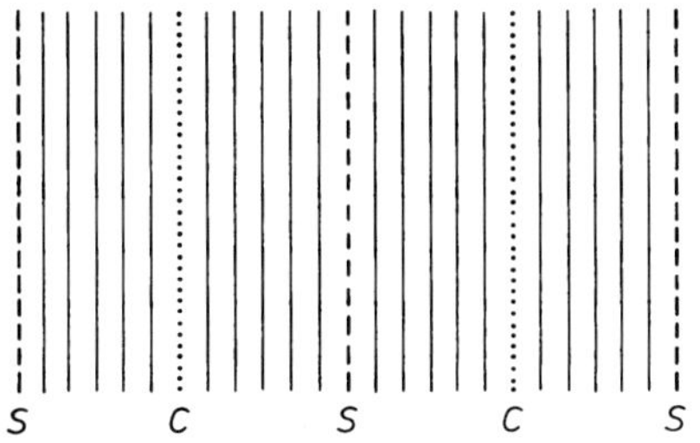

Fig. 8.1. A field rust nursery with spreader rows (*S*) and a check cultivar (*C*)

Spreader rows are usually necessary to ensure that ample spread of the rust occurs. However, they can cause some problems of uneven rust distribution. The rust will be heaviest nearest the spreaders, particularly early in the epidemic. It will tend to even out later, especially if there are susceptible plants throughout the material being tested. Uneven spore loads can be a problem in selecting for some types of resistance such as slow rusting. The resistance can be overwhelmed by heavy spore loads and differences among plants or lines obscured. In special cases such as this, it may be desirable not to have spreader rows but rather to inoculate the whole nursery uniformly.

In setting up a rust nursery, decisions must be made as to whether to use one or more rust species in the same nursery, and whether to use one race or a mixture of races. Although the three rusts tend to interfere with one another, particularly leaf rust and yellow rust, it may be possible to mix stem rust with either leaf rust or yellow rust and obtain readings on each that are adequate for plant breeding purposes. Both leaf rust and yellow rust primarily attack leaves and one will often obscure the other. Of course, if plants are severely attacked by either one they will probably be of no value and will be discarded anyway. At the University of Saskatschewan mixtures of stem and leaf rust have been used successfully for many years.

Whether to use one or a mixture of races depends on the purpose of the test. One race should be used if the purpose is to test for the presence of a specific gene for resistance or to test for resistance to a particularly important race, such as a new, highly virulent one. A mixture should be used if the purpose is to simulate a natural field epidemic or to select for resistance against all of the races that are normally present in the field. In using a mixture, there is no way of knowing what the final composition of the epidemic is, short of actually making collections and identifying them. One or two races may tend to predominate. On the other hand, in many areas it will not be possible to maintain a pure epidemic of one race since natural inoculum will contaminate the nursery. However, if the single race is inoculated early before appreciable amounts of natural inoculum appear, it should predominate.

8.2.1 Field Inoculation Techniques

Several techniques have been used to inoculate field nurseries with urediospores including injection, dusting, spraying, and transplanting infected plants.

8.2.1.1 Injection of a Spore-Water Suspension

The most reliable method of inoculation is to inject a suspension of spores in water directly into the leaf sheaths or stem internodes. Infection is essentially independent of the environmental conditions. A disadvantage of the method is that only the injected internode on injected stems becomes infected.

Spores can be suspended in water by shaking in a bottle or flask. Often a surfactant such as Tween 20 is added to help keep the spores in suspension but Rowell (1984) questions their value. However, at the University of Saskatchewan stem and leaf rust spores germinate poorly in tap water but germinate well if one or two drops of Tween 20 are added per litre. Whatever water source is used, it should be checked to see that the spores germinate normally. Even distilled water, if it has been distilled in copper, may inhibit germination. Hobbs and Merkle (1965) found that using a 10^{-2} M solution of calcium pantothenate increased both the percentage of plants infected and the number of pustules per plant. However, the method has not been adopted widely.

Rowell (1984) describes a procedure that results in spores remaining in suspension. A spore paste is made by adding water one drop at a time to a spore mass in a beaker, and kneading it with a glass rod. When the mixture has the consistency of heavy cream, the bottom of the beaker is placed in the bath of an ultrasonic cleaner for about 1 min, during which additional water is added to make up the desired spore concentration and volume of the suspension. The process removes gas from the spore surface and improves their wettability so that good suspensions can be produced.

When the spore suspension has been prepared, a hypodermic syringe and needle is used to inoculate it into the uppermost leaf sheath. Inoculation is usually done as soon as the stems start to elongate, in order to get the epidemic started as early as possible. The procedure is time-consuming but it is usally sufficient to inoculate one stem every 30 to 50 cm in the spreader rows. If properly done, almost every inoculated stem should be infected. However, at the University of Saskatchewan the procedure has often been less effective with leaf rust than with stem rust. Probably infection with leaf rust requires that the spores be in contact with an emerging leaf blade.

8.2.1.2 Spraying with a Spore-Water Suspension

Water suspensions are seldom sprayed on plants. Under field conditions the spores may start to germinate and then dry out and die unless a dew or rain occurs, or the nursery is irrigated. For inoculating a small nursery, some plants in the spreader rows can be sprayed and then covered with a plastic bag which is anchored at the bottom with soil. This should not be done on a hot, sunny day or the plants may be damaged.

8.2.1.3 Dusting with a Spore-Talc Mixture

One of the simplest procedures for inoculating a sizeable area is to dust the plants with a spore-talc mixture. The ratio of spores to talc is not particularly critical and can vary depending on the quantity of spores available and the area to be covered.

About 1 part of spores to 4 of talc is satisfactory. The critical factor in obtaining infection is the presence of moisture on the leaves following inoculation. Often dusting is done late in the day and spore germination and infection depend on dew deposition. If the humidity is very low, the inoculation can fail. Sometimes the dusting can be done just before or just after a light rain or irrigation. If it is done when the leaves are dry, there is some danger of the spores being washed off by a heavy rain or irrigation. If it is done when the leaves are wet, then conditions must be such that moisture will remain long enough for the spores to germinate and penetrate.

Many types of dusters are available from small, hand-operated to power-operated units. They can be used for any size of nursery.

8.2.1.4 Spraying with a Spore-Oil Suspension

A number of light mineral oils such as Soltrol 170 (Phillips Petroleum Co.) and Mobilsol 100 (Mobil Oil Co.) have proven to be satisfactory carriers for spores and have low phytotoxicity (Rowell 1984). Spores are readily wetted and suspended in oil and this is its major advantage over water. Low concentrations of spores (0.5 mg per ml) are sufficient to give good results. Sprayers that produce a very fine mist are used. For larger areas, several types of hand- or power-operated low volume sprayers are available. Oil suspensions work well for greenhouse and growth chamber inoculations using humidity chambers. Results have varied considerably in the field. Rowell and Hayden (1956) successfully used oil suspensions for field inoculations. However, at the University of Saskatchewan almost no infections have been obtained. The difference probably depends on the availability of moisture on the leaves. Dews are often light at Saskatoon and may not be sufficient to wet spores that have been suspended in oil.

8.2.1.5 Using Infected Plants

For relatively small nurseries, it is possible to start plants in the greenhouse or growth chamber, infect them with rust and move them to the nursery. Pots of infected plants may be simply moved to the nursery. If watered adequately, they should produce spores for up to 3 or 4 weeks. Alternatively, the plants may be transplanted. In this case, they will transplant most easily if they are in peat pots or some other type of container from which they do not have to be removed. Frequent watering will still be required to ensure good establishment.

8.2.2 Measuring Rust Severity

In most wheat breeding programs, it is sufficient to decide the degree of resistance that is acceptable and select only plants with at least that level of resistance. For other studies, more detailed readings of rust severity are often required. A common procedure is to record the percentage of the leaf or stem that is covered with uredia. The leaves or stems are compared with diagrams on which various percentages of the area have been covered with spots of various sizes representing pustules. Even when all the leaf or stem is invaded by mycelium only about 37% of the surface is covered with pustules and this is taken as 100% severity. The first scale was developed by N. A. Cobb and then modified by Petersen et al. (1948) (Fig. 3.2). The readings are fairly subjective and are particularly difficult at the upper end of the scale. Most workers do not attempt to rate the plants into classes smaller than 10%, except at the bottom end of the scale. For example, classes might be 0–1, 5, 10, 20, 30, 40, 50, 60, 70, 80, 90, and 100%. Because of the difficulty in distinguishing severities at the upper end, for many purposes the classes can be reduced to 0–1, 5, 10, 20, 40, 70, and 100%. Different workers will undoubtedly vary in the ratings they give the same plants and one person may rate the same plants somewhat differently at different times. When the ratings for different lots of material are to be compared, it is important that they be read under consistent light conditions. Despite these difficulties, the procedure can provide very useful data.

The scales do not work particularly well for yellow rust because the pustules are long stripes rather than discrete pustules.

Often plants are rated for infection type as well as rust severity using the following classes:

O – no visible infection.
R – resistant – yellow, chlorotic or necrotic areas which may contain small uredia.
MR – moderately resistant – small pustules surrounded by chlorotic or necrotic areas.
M – mesothetic – pustules of a complete range of sizes with some chlorosis and necrosis.
MS – moderately susceptible – moderate sized pustules with some chlorosis.
S – susceptible – large pustules with little chlorosis.

Thus, readings might be 5R (5% severity, resistant infection type), or 60S (60% severity, susceptible infection type). Reading infection types can be difficult for stem rust on adult plants. Often the uredia are of different sizes on different parts of the plant, larger on basal internodes and particularly just above the nodes, but smaller on the peduncles. However, this is not a mesothetic reaction in which uredia of different sizes occur mixed together.

In international tests, it has become common practice to combine severity and infection type to give a coefficient of infection. The infection types are given values: 0 = 0, R = 0.2, MR = 0.4, M = 0.6, MS = 0.8, and S = 1.0. A reading of 10MS gives a coefficient of infection of 10 × 0.8 = 8.0. The coefficients for a cultivar or line can be averaged over different tests. Because the stage of growth can have a considerable effect on rust severity, some workers record the stage when the readings are taken. This is particularly important if the lines being tested vary considerably in maturity.

When the progress of a rust epidemic has been measured at several times during its development, the rate of rust development (r) and the area under the disease progress curve (AUDPC) can be calculated. The former can be calculated for any two times of reading from the formula (Vanderplank 1963):

$$r = \frac{2.3}{t_2 - t_1} \log 10 \frac{x_2(1-x_1)}{x_1(1-x_2)} ,$$

where x_1 is the disease severity at time $1(t_1)$ and x_2 at t_2. The area under the disease progress curve can be calculated as:

$$AUDPC = \sum_{i=1}^{k} \frac{1}{2}(s_i + s_{i-1})$$

s_i = rust severity at the end of period i and k = the number of times rust readings were taken (Wilcoxson et al. 1975). Both parameters can be used to compare rust development on different genotypes but the AUDPC is probably more reliable.

8.2.3 Dangers from Artificially Inoculated Field Rust Nurseries

Artificially inoculated field rust nurseries pose two threats:

1. If a nursery is inoculated either before the arrival of natural inoculum or in an area where there is no natural inoculum, the rust may spread to nearby commercial fields and cause damage. The threat is greatest if the rust season is long and an epidemic can build up over a long period. However, it is greatly diminished if the nursery is even 1 or 2 km from the nearest commercial wheat. It takes time for an epidemic to build up in a nursery and produce appreciable amounts of spores. Considerable dilution will occur when the spores are blown 1 or 2 km to the nearest field. It will then take several generations for the rust to build up in the field and serious damage is unlikely to occur. Nevertheless, the concerns of local farmers and politicians should be considered. Often the best defense is to make sure that they know what is being done and the great benefits to farmers that come from breeding for resistance.
2. If a highly virulent race is used that is not present in nature, it may escape and become a threat to commercial crops. As a general rule, virulent races that are not present in an area should not be imported from some other area. If testing with a special race is essential, international cooperation is such that it can usually be arranged with another scientist in an area where the race is already present.

If proper precautions are taken, the danger from rust nurseries is minimal. The potential benefits far outweigh the risks.

8.3 Greenhouse or Growth Chamber Rust Tests

Most wheat breeding programs do not have enough greenhouse or growth chamber space to handle large volumes of plant breeding material. Nevertheless, greenhouse or growth chamber tests can be used for limited amounts of material when it is particularly important to test with a specific race or to speed up the breeding process. Testing can often be done in the off-season. It is useful to know something about the inheritance of resistance in the parents being used. An important advantage of greenhouse or growth chamber tests is that pure rust cultures of known genotype can be used and provide information about the genotypes of the parents of the crosses being tested.

Seedling tests often start with the F_2 generation. Seedlings are rusted, resistant plants are saved and grown to maturity. Sometimes resistant plants are transplanted to the field, where selection for agronomic characters can be done. Transplanting is easier if the material has been planted in peat pots or some other units that can be planted directly into the field without disturbing the plants. Even paper cylinders will work. Since resistance is usually dominant, further testing will be necessary to select homozygous resistant families or lines. When this is done will depend on the schedule of field and greenhouse testing.

Alternatively, field testing for agronomic characters can be done first and seedling rust tests in greenhouses can be carried out on reasonably homozygous lines. This is particularly useful if field rust testing is difficult and not always successful. If desired, the same line can be tested with two or more rusts, or two or more races of the same rust.

Some genes condition resistance that is effective only in adult plants. Greenhouse or growth chamber tests can sometimes be used to identify these genes but it is essential to know that they will be expressed. For example, the effect of *Sr2* is difficult or impossible to detect in greenhouse tests of adult plants with stem rust, and adult plants of the Canadian cultivar Manitou are susceptible in the greenhouse to leaf rust races to which they are resistant in the field. However, slow rusting can often be evaluated as easily in a greenhouse or growth chamber as in the field.

8.4 The Pedigree System Of Breeding

A standard pedigree system has been used in numerous breeding programs for rust resistance. While many modifications of the system are possible, the following outline is representative of the procedures used.

1. The parents. The choice of parents is extremely important. They must carry all of the genes that are desired in the progeny. At the very least, one parent must be resistant to prevalent races of each rust that is a problem. If the objective is to obtain segregates that are more resistant than either parent, then the parents must carry different genes for resistance. Multiple crosses may be used to combine resistance from several parents. If it is hoped to produce a commercial cultivar directly from a cross, then the parents must be reasonably well adapted to the area and carry other desired characters. To use genes from an unadapted parent, it is usually necessary either to backcross it once or twice to an adapted cultivar or to use it in a three-way cross with two adapted parents.

2. The crosses. The parents can be grown and the crosses made in the field or in growth chambers or greenhouses, whichever is convenient. Only limited numbers of hybrid seeds are needed for single crosses but more should be produced for three-way crosses, double crosses, and backcrosses, since the F_1 plants will segregate.

3. The F_1 hybrids. The F_1 plants should be grown under favorable conditions to ensure a good seed increase. It is probably not a good idea to grow them in a rust nursery in case they become rusted heavily enough to severely reduce the quality of the seed. Seedlings can be rusted to determine whether resistance is dominant or recessive. The F_1 plants from single crosses should be examined for any off-types and

particularly for selfs. Three-way, double cross and backcross F_1s will segregate for various characters and some selection may be done in them. The F_1 plants can be harvested and threshed separately and individual F_2 families grown, but often the plants are handled in bulk, particularly from single crosses.

4. The F_2 generation. The F_2 populations are grown as spaced plants, 15 to 30 cm apart, in a rust nursery. Resistant plants having other desirable characters are selected. Selection can be done in various ways depending on the rust and the frequency of resistant plants. Selection for leaf rust or yellow rust must be done during the growing season before the leaves senesce. Usually it is necessary to make selections more than once during the season, particularly if there is much segregation for maturity in the crosses. Some types of resistance, such as slow rusting, may be more easily identified at early stages in the epidemic. Later the resistance may be overwhelmed by heavy spore loads, even though it would provide adequate resistance in commercial fields where spore loads would be light. If only a limited number of F_2 plants are susceptible, they can be pulled or broken over. If many plants are susceptible, then resistant plants should be marked with tags or a spot of spray paint. With stem rust the selections can be made at maturity, although it is sometimes helpful to examine the severity of leaf infections and mark resistant plants. At maturity, agronomically desirable rust resistant plants are pulled or a single head is taken from each. The plants or heads are threshed separately.

5. The F_3 generation. Individual F_3 plant or head progenies of 30 to 50 plants are grown in short rows in a rust nursery. Selection is done in two steps. First, rows that are either rust-resistant or at least segregating, and appear agronomically desirable, are selected. Then plants are selected within these rows. Again, either whole plants or individual heads may be harvested and threshed separately.

6. The F_4 generation. The F_4 families are handled much like the F_3 families. However, because there is less segregation within families, more emphasis is placed on selection among families and less on selection among plants within families. Occasionally, a very promising, uniform row can be harvested in bulk and entered into the system for evaluating agronomic and quality characteristics. Usually, individual plants or heads are harvested.

7. The F_5 generation. By this stage families should be fairly homozygous and uniform. They can be planted more thickly and many breeders harvest selected rows in bulk at this stage. Others prefer to again harvest individual plants or heads and not start bulking individual rows until the F_6 generation.

8. Homozygous lines. Although there will always be some residual heterozygosity, most F_6 or F_7 lines will be reasonably uniform and can be entered into the normal system for testing agronomic and quality characteristics. Usually this will begin with a fairly limited yield test probably at one location. As the poorer lines are eliminated, yield testing is expanded to cover the potential area of production. More detailed rust testing can be done at this stage, including tests with specific races either in field nurseries or more probably in greenhouses or growth chambers. These are valuable in determining the spectrum of resistance carried by potential new cultivars. New, particularly virulent races that are not yet widespread can be used to determine the probable durability of a line.

The name "pedigree system" refers to the fact that pedigrees of each line are maintained during the breeding process. Each cross is designated by a number, a letter or a cross abbreviation, e.g., a Thatcher/Selkirk cross could be designated Tc/Sk. If a bulk F_2 population is grown, each selected plant is numbered and the F_3 families are designated Tc/Sk-1 F_3, Tc/Sk-2 F_3, etc. Each generation a new number is added, e.g., if two plants are selected from Tc/Sk-2 F_3, their progenies would be labeled Tc/Sk-2-1 F_4 and Tc/Sk-2-2 F_4. Each generation a field book is prepared and important notes are taken on the families or lines. At any stage, a line can be traced back through all previous generations and the notes examined.

The basic pedigree system can be modified in various ways to suit specific objectives, particular situations and available resources.

Several wheat breeders have published details of their breeding systems. For example, at the Agriculture Canada Research Station at Winnipeg a typical flow sheet is as follows (from Green and Campbell 1979):

"Year 1 – Cross made in growth cabinet. F_1 grown in Mexico.

Year 2 – F_2 grown in rust nursery at Winnipeg. Selection for resistance. F_3 headrows grown in Mexico. No planned selection.

Year 3 – F_3 headrows grown in Winnipeg rust nursery. Selection for resistance and other attributes. Quality test after harvest. F_5 headrows grown in Mexico. Lines with poor quality in F_4 discarded. No planned selection.

Year 4 – F_6 headrows grown at Winnipeg. Selection for a wide range of characters. Quality test after harvest. F_7 headrows grown in New Zealand. Lines with poor quality in F_6 discarded. An increase generation.

Year 5 – A 5 m increase row grown at Winnipeg and rust nursery test."

For the next 5 years, the selected lines are given successively expanded yield tests. In each year they are tested in a field rust nursery with a mixture of virulent races, and with individual races as seedlings in the greenhouse.

At the Plant Breeding Institute, Cambridge, England, the wheat breeders use a pedigree system with intensive selection and early generation yield and quality testing (Bingham and Lupton 1987). About 2000 F_2 plants are grown from a cross and on average about 50 are selected on the basis of resistance to diseases such as leaf and yellow rust and agronomic characters. One ear is harvested from each plant unless emphasis is to be placed on bread-making quality, in which case the whole plant is harvested. Seed from such plants is ground and tested by near-infrared (NIR) analysis for grinding resistance, milling texture, and protein content, by the sodium dodecyl sulfate (SDS) sedimentation test for protein strength, and chemically for alpha-amylase activity. For winter wheat the F_3 lines have to be planted before the quality tests have been done, so quality selection is delayed a generation. A 1-m F_3 row is grown from each selected F_2 plant. Intensive selection is practiced among F_3 rows and 12 ears are harvested from each selected row. The F_4 is handled in the same ways as the F_3. Selected F_5 lines are grown in a single-plot yield test with a check cultivar after every five plots, as well as in short rows for further selection. Plant-to-row selection continues until uniformity has been achieved. Yield testing is expanded in F_6 and F_7 and further quality testing is done. Selected lines are then entered in national trials and seed multiplication begins.

8.5 Off-Season Nurseries

As outlined above, the pedigree system involves only one generation per year. Wheat breeders are always interested in speeding up the process. Off-season nurseries are frequently grown in order to produce two generations a year. In more tropical areas they can sometimes be grown at the same location as regular season nurseries, perhaps using irrigation. In other cases it may be necessary to go to another part of the country or even to another country. Often it is not possible to set up rust nurseries for the off-season crop so selection for rust resistance occurs only every second generation.

8.6 The Bulk System of Breeding

In a true bulk system, the early generations of a cross are planted in bulk at normal seeding rates and the material is allowed to evolve through natural selection. No artificial selection is applied. To breed for rust resistance, the system can be modified by growing the bulks in a rust nursery. An early, heavy epidemic of rust will cause shriveling of seed produced on susceptible plants. When a bulk sample of seed is planted in the next generation, the shrunken seed should have reduced viability and even those that are viable should produce weaker, less competitive plants. The system can undoubtedly be made more efficient by cleaning out as much of the shrunken seed as possible with appropriate sieves and wind separation. A gravity table can also be used to select high density seed. The speed of evolution toward resistance will depend on the severity of the rust epidemics. When sufficient progress has been made and individual plants are reasonably homozygous, they are harvested and threshed separately. Individual plant progenies are grown in a rust nursery and checked for resistance. Homozygous rows that have desirable agronomic characteristics are bulked and entered into the yield testing system.

McFadden (1930) used several types of screening and air cleaning to select for short, plump F_4 seeds from a cross between Yaroslav emmer and Marquis bread wheat that had been subjected to a heavy rust epidemic. The method was successful in selecting plump, bread wheat type kernels that came from disease resistant plants. Tiyawalee and Frey (1970) used the air cleaning on a plot thresher to remove shrunken kernels from an oat population. The frequency of resistance to crown rust, increased from 0.21 in the F3 to about 0.35 in the F10. Thus, such a system can be effective.

8.7 A Modified Pedigree System

The pedigree system involves a considerable amount of labor in harvesting and threshing separate plants or heads, planting large numbers of short rows, taking notes and keeping records. Most, if not all, plant breeders have a limited amount of land and labor available. If large amounts of resources are used in early generations, less will be available for yield testing. Breeders must decide which stage is most important and strike a balance. At least for yield it can be argued that selection among homozygous lines is more effective than selection in early generations.

In the durum wheat breeding program at the University of Saskatchewan, a system is used that minimizes the effort involved in early generations while maximizing the resources available for yield tests. The F_1 and F_3 generations are grown in bulk in off-season nurseries in New Zealand with no selection. The F_2 and F_4 generations are grown at Saskatoon in long rows of spaced plants in a rust nursery inoculated with a mixture of the most prevalent and virulent leaf and stem rust races in North America. At least twice, after the rusts are spreading well, technicians walk through the nursery and break over or pull susceptible plants. In most crosses the number of susceptible plants is not large. For the F_2 generation, at maturity the breeder goes through each cross and harvests one head from each agronomically desirable plant that is resistant to stem rust. All the heads from one cross are put in a bag, threshed in bulk, and a bulk sample planted for the F_3 generation in New Zealand. In New Zealand, a single head is harvested from all but the very poorest plants, bulked, and a bulk F_4 sample planted at Saskatoon. The F_4 is handled in the same way as the F_2, except that three heads are taken from each selected plant to provide ample seed for an F_5 row. The three heads are stapled together with a tag around their peduncles. Each set of three is threshed separately and an F_5 row planted from it in a rust nursery. Every 12th row is planted to a commercial check, two cultivars being used alternately. The F_5 rows are examined to see that they are homozygous for leaf and stem rust resistance (yellow rust is not a problem). Rows that look promising in comparison with the checks are harvested in bulk and tested for protein and pigment concentration, and for SDS sedimentation value. The final selections are entered into the regular yield testing system.

The procedure greatly reduces the space and labor required for early generations while allowing selection for rust resistance and other characters. As a result, more lines can be handled in yield tests and 600–700 new lines are entered each year.

8.8 Backcrossing

The transfer of genes for resistance into a cultivar by backcrossing is particularly useful when a cultivar is available that lacks rust resistance but is otherwise very desirable. For example, an excellent cultivar may become susceptible to a new, virulent rust race. While it is not absolutely essential to know the inheritance of resistance in a donor, it makes the planning of the backcrossing much easier.

Genes for specific rust resistance are often dominant and can easily be backcrossed into a recurrent parent. Backcrossing can be done in a greenhouse or growth chamber and the seedlings tested with an appropriate race each generation. Even recessive genes can be handled easily. A cross and a backcross can be made and the progeny selfed to recover the resistance. Two more backcrosses can be made and the process repeated. If speed is essential, the generations of selfing can be eliminated by progeny testing. After each backcross, a number of plants are both selfed and backcrossed again. The plants must be numbered and the identity of the selfs and backcrosses maintained. The progenies from the selfed heads are tested to determine which parent plants were heterozygous and which homozygous susceptible. Backcrosses made on the heterozygotes are saved and the process repeated.

Occasionally some loss of resistance occurs during backcrossing, probably as a result of the loss of modifying genes. Usually this is not a major problem. However, if it is important to maintain full resistance, then stronger selection must be applied during the backcrossing. After the first backcross and every two additional backcrosses the plants can be selfed and selection made for the full resistance.

Genes for specific resistance that is effective only in adult plants can be more difficult to transfer. Depending on the stage at which the resistance becomes effective, it may be difficult or impossible to determine whether plants are resistant in time to make crosses. If this is the case, crossing has to be delayed until the next generation. Some genes for adult plant resistance, such as *Sr2*, are difficult to identify in greenhouse tests. In such cases, field tests must be used.

Resistance that is quantitative and controlled by several genes is much more difficult to transfer by backcrossing but it can be done. A cross and a backcross are made. Then the material must be selfed and selection carried out for several generations until the resistance has been recovered. This can be slow, particularly if resistance can be identified only in field tests of reasonably homozygous lines, as is often the case for slow rusting. A large enough number of lines must be tested to ensure that the desired phenotype will be obtained. For example, if four genes are involved and a heterozygous F_1 is backcrossed to the recurrent parent, only one plant in 16 will carry all four genes. And, of course, its progeny will segregate in the next generation. Once resistance has been recovered, two further backcrosses can be made, followed again by selfing and selection.

The number of backcrosses that are required depends on how completely the recurrent parent must be reconstituted. Two or three backcrosses are usually sufficient to produce lines that are morphologically very similar to the recurrent parent. Selecting for the phenotype of the recurrent parent in the early backcross generations will speed up the process. If it is not essential to completely reconstitute the recurrent parent, then two or three backcrosses should be sufficient. The fewer backcrosses that are made, the better is the chance of selecting for improvement in characters other than rust resistance. If it is important to essentially reconstitute the recurrent parent, then five backcrosses are usually sufficient. At that point the backcrosses should carry 98.4% of the genes of the recurrent parent that are not linked to the locus involved in the rust resistance. In the production of Stewart 63 durum wheat (Ethiopia St 464/8* Stewart) at the University of Saskatchewan, homozygous lines were first produced after five backcrosses. However, none of the lines was equal to Stewart in pasta-making quality. After two additional backcrosses an acceptable line was obtained (Knott 1963). It is not known whether the deleterious effect on quality was due to genes from Ethiopia St 464 that were still present by chance after five backcrosses, or whether one or more linkages were broken during the two additional backcrosses.

No wheat cultivar is 100% homogeneous and many include considerable variability. For this reason it is essential to use a representative sample of plants of the recurrent parent during backcrossing. It is particularly important that a number of plants be used in the final backcross and that a number of homozygous lines be produced and mixed to form the final product. Otherwise there is danger of an atypical line being produced that will not be equal to the recurrent parent in some characters.

Backcrossing is a conservative procedure. The basic objective is to change one character in an otherwise satisfactory cultivar. However, by the time the backcross line has been produced, the recurrent parent may be obsolete. In areas where rapid progress is being made, cultivars become obsolete quickly. However, in some areas such as the Canadian prairies where the climate limits yields and breeders must meet exacting requirements for bread-making quality, improvement in the yield of cultivars is relatively slow. As a result, Dr. A. B. Campbell and his colleagues at the Agriculture Canada Research Station in Winnipeg have made very extensive and successful use of backcrossing. Many of the current Canadian cultivars are backcross derivatives of Thatcher. Although they are backcross derivatives, there have been some improvements in characters other than those being transferred, including yield.

8.9 Partial or Incomplete Backcrossing

The problem of an obsolete recurrent parent can be partially overcome by using only one or two backcrosses. The material should then have 75 or 87.5% of the genes of the recurrent parent, if linkage to the genes being transferred is ignored. However, there will still be enough genes from the donor parent to allow selection for characters other than rust resistance and, thus, to produce lines that are better than the recurrent parent. Clearly the procedure will work best if the donor parent is reasonably well adapted and carries desirable genes for characters other than rust resistance.

The CIMMYT wheat breeding program in Mexico makes extensive use of partial backcrossing for a slightly different reason. The objective of the program is to put genes for disease resistance into desirable genetic backgrounds. The material is then sent to many countries for use in their programs. It is desirable for the material to retain sufficient variability that the breeders in these countries can select out lines that are adapted to their particular conditions.

8.10 Convergent Backcrossing

A second problem with backcrossing is that normally only one character can be improved at a time, while most cultivars have more than one fault. This can be overcome by a procedure called convergent backcrossing. Separate backcross programs with one recurrent parent are carried on simultaneously, each program improving one of a number of desirable characters, e.g., resistance to two or more pathogens, different resistances to one pathogen, resistance to sprouting, reduced height, etc. At the end of the backcrossing programs the various lines can be intercrossed to combine several of the characters in one final product. The process can be slow and the danger is increased that the recurrent parent will have become obsolete. On the other hand, the recurrent parent should have been substantially improved.

8.11 Single Seed Descent (SSD)

A few wheat breeders are using a procedure called single seed descent to speed up the production of relatively homozygous lines. Normally the procedure starts at the F_2 generation. Single seeds are taken from plants of an F_2 population and sown, usually in a bulk plot. The process is repeated for as many generations as desired, i.e., until the desired level of homozygosity has been reached. Individual plant progenies are then grown and tested for various characters. Since germination and survival is never 100%, the size of the population will decline each generation. This must be taken into account in deciding the size of the F_2 population with which to start. The loss can be largely prevented by planting two seeds from each plant in each generation and thinning to one plant. However, this requires considerable extra labor in harvesting two seeds per plant and keeping the pairs separate, sowing pairs of seeds, and thinning.

In the standard SSD procedure, selection is not practiced in the early generations. In fact, in some cases the objective is to produce a random set of near-homozygous lines for use in genetic studies. However, an F_2 population could be grown in a rust nursery, selections made for rust resistance and other characters, and a single seed taken from the selected plants to start the SSD procedure. The procedure can be modified to suit individual needs.

The main objective of the SSD procedure is to produce populations of reasonably homozygous lines as quickly and with as little labor as possible. The lines can then be tested for characters such as yield that are not readily measured in individual plants. Since only a single seed per plant is required, the plants are often grown in a greenhouse or growth chamber under very crowded conditions that result in small plants and a short generation time (Grafius 1965; Tee and Qualset 1975). Procedures using sand cultures, watered once or twice with a nutrient solution, have been successfully used. Under such a system most plants will produce one head and a few seeds. However, some plants will be sterile. Semidwarf plants may tend to be shaded out by taller plants and competition can change the composition of the population. The procedure is then not true SSD. However, it should be adequate for breeding purposes.

Allan (1987) describes a procedure using single spikes which has some advantages over the use of single seeds. One spike is harvested from each F_2 plant. About 10–30 seeds from each spike are planted in hill plots in the field or in greenhouse pots. A single spike is harvested from each plot or pot to produce the next generation. The process is repeated until the desired degree of homozygosity has been reached. The procedure requires more space than single seed descent but there is little danger of losing lines. In wheat breeding, where the objective is not to produce a random sample of lines, selection can be done among hill plots.

8.12 Recurrent Selection

Recurrent selection can take many forms. Basically it involves selecting a number of parents, making all possible crosses among them, or at least as many crosses as is feasible, growing out the progeny, selecting for the desired character, and then intercrossing the selected material to start another cycle. The objective is to have repeated cycles of recombination followed by selection, thus concentrating the desired genes. Recurrent selection has been used very successfully in the breeding of many cross-pollinated crops. It has been used only occasionally in self-pollinated crops because of the difficulty in many of them of producing the considerable number of crosses required.

Nevertheless, recurrent selection has been used for a number of characters in wheat such as protein content (McNeal et al. 1978; Löffler et al. 1983), heading date (Avey et al. 1982), kernel weight (Busch and Kofoid 1982), and yellow rust resistance (Sharp 1979). For rust resistance, it will generally be most useful for resistance that is controlled by several genes each having small effects. Such resistance is often expressed best in adult plants and can be masked by major genes for specific resistance. This masking effect can be prevented in either of two ways. One possibility is to select parents that lack major genes for specific resistance to the race used. Alternatively, the segregating progeny of the first crosses can be tested at the seedling stage and all *resistant* plants removed. This should eliminate dominant genes for specific resistance. If recessive genes are present, two generations of selection for seeding susceptibility will be necessary. The most virulent race available should be used so that the parents carry as few effective genes for specific resistance as possible (see Sect. 8.15.2 for more detail). Several generations of selection among adult plants may be necessary to recover resistant plants for the second cycle of crossing. Because of the difficulty of making all possible crosses among, for example, 20 parents (190 crosses), it should be sufficient to use each parent in four or five randomly chosen crosses. Each parent will be adequately represented in the next generation and there will be ample opportunity for recombination.

Either genetic or cytoplasmic male sterility can be used in wheat to facilitate cross-pollination and obviate the need for hand-made crosses in a recurrent selection program. The procedure is simplest using a dominant gene for male sterility. It may be desirable to start by backcrossing the gene for male sterility into an adapted parent. The following procedure can then be used:

1. Mix seed of the male-sterile line and the parents selected to start the recurrent selection program. Plant a bulk plot, mark male-sterile plants at anthesis and harvest seed from them. The seed should all be hybrid.

2. Grow a bulk plot in a rust nursery and mark the male-sterile plants (50% should be male sterile). If rust resistance can be determined before heading, remove susceptible plants so that crossing occurs only among resistant plants. Otherwise, mark resistant plants as resistance becomes apparent. Harvest seed from resistant, male-sterile plants. Only very moderate levels of resistance may be present in early generations.

3. Repeat the procedure in (2) as often as necessary until adequate levels of resistance have been obtained.

4. In the final generation, select resistant, fertile plants which will lack the gene for male sterility, and grow out progeny rows to select for homozygosity. To prevent outcrossing to undesirable material, the recurrent selection plots should be as isolated as possible from other wheat plots.

Similar schemes can be worked out for either recessive genes for male sterility or cytoplasmic male sterility and fertility-restoring genes.

With the development of chemical hybridizing agents (CHAs), these can be used to promote outcrossing in a recurrent selection program. The procedure is simpler than using either genetic or cytoplasmic male sterility. The material is planted in rows and alternate rows are sprayed with a CHA. If rust reactions can be determined before heading, susceptible plants are pulled so that crossing occurs only among resistant plants. Plants are harvested only from the sprayed, male-sterile rows. If rust reactions cannot be determined until after heading, resistant plants in the sprayed rows are marked and harvested at maturity. The procedure can be repeated as often as necessary. After the final cycle, progeny rows can be grown out and selected for homozygosity.

8.13 Using Knowledge About the Genetics of Resistance

So far in this chapter, little has been said about the genetics of the resistance being used. In many cases breeders do not know much about the genes in the parents that they are using, only that the parents are resistant. A good example is the situation with durum wheats in North America. Most cultivars have acceptable resistance to leaf rust but little research has been done on its inheritance. In many crosses for durum wheat breeding, little segregation occurs. This may mean that there are relatively few genes for resistance and many parents carry genes in common. Alternatively, many genes may be involved, most parents may carry several, and most progeny of crosses receive some genes and are resistant.

It can be dangerous to breed for resistance without any knowledge of the genes carried by the parents. Selection may favor one particularly effective gene so that many cultivars end up carrying it. As a result, wheat production in an area will be particularly vulnerable to the development of a new race of the rust that is virulent on the gene. Such a situation occurred recently with triticale in Australia (McIntosh et al. 1983). A new race of stem rust was discovered that was virulent on a number of cultivars and it was found that many triticales depended on *Sr27* for resistance. The gene *Sr27* is carried on a rye chromosome or a wheat-rye translocation. In wheat, some genes are found in a large number of cultivars, for example, *Sr2, Sr5, Lr10, Lr13,* and *Yr2,* and most are no longer of any value by themselves, although they may be useful when combined with other genes.

Much of the early breeding for rust resistance in wheat involved producing cultivars with a new gene for resistance each time resistance in existing cultivars was overcome by a new race. The stem rust situation in Australia provides an excellent example because of the speed with which new races appeared and the rapid response of the breeders. As the genes *Sr6, Sr11, Sr9b, SrTt1(Sr36),* and *Sr17* were used in cultivars they were overcome by the rust (Luig 1983).

8.13.1 Pyramiding of Genes

The short lifespan of single genes for stem rust resistance led the Australian breeders and others to produce cultivars that combined several effective genes. The procedure is often called gene pyramiding. When breeders put together several genes for which the prevalent races of the pathogen are avirulent, the development of a virulent genotype of the rust requires the combination of all of the required mutations for virulence. The chance of two or more mutations occurring simultaneously in the pathogen is small. Unfortunately, what often happens is that cultivars are released that carry the genes singly. The rust can then use these cultivars as stepping stones for step-wise mutations overcoming one gene at a time. Eventually a genotype virulent on all of the genes carried in a complex cultivar is produced. This occurred with the Australian cultivars, Gamut (*Sr6, Sr9b, Sr11, SrGt*), Mendos (*Sr11, SrTt1, Sr17, Sr7a*), and Oxley (*Sr5, Sr6, Sr8a, Sr12*) (Luig 1983). However, the four cultivars, Timgalen, Timson, Cook and Shortim, each of which carries several genes for resistance, have remained resistant. All four carry *Sr6* and *SrTt1(Sr36)*. Although virulence on each gene has occurred in separate rust isolates in the field, no isolate combining the two virulences has been found. However, several isolates combining the two virulences have been produced in the laboratory (Luig 1983). The release of cultivars with diverse genotypes for rust resistance has undoubtedly helped in the control of stem rust in Australia.

In the pyramiding of genes it is desirable to know which genes to combine and which genes are carried by various parents. In crosses it may not always be easy to identify plants or lines carrying the required gene combination. Sometimes genes can be identified by using rust isolates to which one gene conditions resistance but the others do not. If two genes both condition resistance but not immunity to all available races, it may be possible to identify the gene combination by looking for increased resistance. This is particularly likely to be identifiable in heavy field epidemics of the rust. If one gene is temperature sensitive, tests for a second gene can be done at a temperature that prevents the expression of the first gene. Occasionally gene linkages can be used to follow a gene for resistance. For example, *Lr24* and *Sr24*, two genes derived from *Agropyron elongatum*, are inherited as a unit. The gene *Sr24* could be combined with other genes for stem rust resistance simply by testing for leaf rust resistance (assuming that the other parent does not also carry leaf rust resistance), or vice versa. The more information that is known about genes that are to be combined, the easier it is to devise a strategy to combine them. If all else fails, it may be necessary to take homozygous resistant lines from a cross, cross them to a susceptible line and determine the number of genes that are segregating.

8.13.2 Anticipatory Breeding

The rust pathogens are constantly evolving. Often the wheat breeder can anticipate the direction in which the pathogen will evolve in the field on the basis of two types of evidence.

1. If extensive race surveys are carried out, the occurrence of a new race can be detected before it becomes prevalent. Trap nurseries containing a set of resistant lines carrying various genes and combinations of genes can be planted throughout

the wheat-growing area of a country. Observations on these nurseries can be used to determine changes that are occurring in the rust population and to detect the presence of new, virulent races that are a threat.

2. Much of the evolution of the pathogen occurs because of selection pressure resulting from genes carried by the cultivars that are being grown. When cultivars are released that carry a gene not previously present, the breeder can anticipate the development of a race that can overcome the resistance. By regular surveys of new, resistant cultivars in the field, the breeder can detect new, virulent races, isolate them, and use them in breeding.

Using information of this sort, the breeder can anticipate changes in the rust pathogens and breed resistant cultivars before they are actually needed. New and potentially dangerous isolates are used first to identify sources of resistance and then to breed resistant cultivars.

Wheat scientists at the University of Sydney have carried the process a step further. When a new, resistant cultivar is released, instead of waiting for a virulent mutant to occur in the field, they have used chemical mutagens to produce a virulent genotype of the pathogen (Luig 1979). This genotype can then be used to start breeding for the next cycle of resistant cultivars. However, it may not always be possible to produce a virulent mutant, since Luig (1979) found tremendous variability in the rate at which different genes for avirulence mutated to virulence. Despite considerable effort, no mutants for virulence on *Sr26* were found.

The use of highly virulent rust genotypes in any form of anticipatory breeding involves some danger. Great care must be exercised to ensure that laboratory genotypes that are not present in the field do not escape. There may even be some danger if naturally occurring rust genotypes are used to inoculate a rust nursery and then cause early infections in nearby wheat fields. At the University of Saskatchewan, the rust nurseries are usually about 3 km from the nearest commercial wheat field. Even if a few spores did reach a field, the crop would mature before any significant rust build-up occurred. In areas where the growing season is much longer, the spread of rust from artificially inoculated nurseries could be much more of a threat.

8.14 Durable Resistance

More often than not the use of genes for specific resistance to the rusts has resulted in rapid evolution of the rust pathogen and a breakdown of resistance. As a result, wheat breeders have become much more interested in types of resistance that might be more durable – nonspecific resistance (if it exists), slow rusting, partial resistance, adult plant resistance, etc.

In breeding for durable resistance, two questions immediately arise: (1) How can durable resistance be recognized? (2) How can it be used in a breeding program? The fact that resistance involves slow rusting, or partial resistance, or adult plant resistance is no proof that the resistance will be durable. Genes that condition intermediate levels of specific resistance will often appear to cause slow rusting or partial resistance. Several of the known genes for rust resistance act only in adult plants but are specific.

8.14.1 Identifying Durable Resistance

Durable resistance often seems to be controlled by several genes, each having a small effect (polygenic resistance). Such resistance may be durable because of the obstacles it places in the way of the evolution of virulence in the pathogen. With the most extreme type of specific resistance, a cultivar with a gene for immunity, the selection pressure is maximal – the rust pathogen either carries a gene for virulence or it does not survive. A virulent mutant has no competition and increases rapidly. On the other hand, polygenic resistance is often incomplete and, therefore, places less selection pressure on an avirulent culture which can reproduce to some extent. Furthermore, if a mutant occurs that overcomes one of the polygenes, it will have only a small selective advantage and will build up slowly in competition with other genotypes. A rust genotype that can overcome several genes may take a long time to appear. If the resistance is truly nonspecific, the pathogen may gradually build up aggressiveness on it. However, there is little evidence to indicate how far such a process can go.

Clearly, genetic studies will be helpful in identifying durable resistance. However, the final test must always be the growing of a cultivar on a large scale for a number of years in an environment favorable for rust development (Johnson 1981). Widespread testing of a cultivar in small-scale plots is not adequate. It will test for the presence of a virulent genotype in the current rust population, but will provide no idea of the likelihood of the development of a virulent genotype in the future when large areas are planted to the resistant cultivar.

Various ways of recognizing nonspecific resistance have been suggested. For example, look for genotypes that show slow rusting – susceptible-type pustules but a slow build-up of the rust. Slow rusting can be measured by taking several rust readings, a minimum of three, and drawing disease progress curves (Sect. 8.2.2). These can then be used to calculate the area under the disease progress curve (AUDPC). It must be remembered, however, that genes for specific resistance that condition intermediate infection types can also have small AUDPC's. In theory, genotypes carrying them should show a normal number of smaller pustules compared to slow rusting lines that show fewer pustules of normal size. However, adult plants often show a range of pustule sizes and the distinction may not be clear.

The wheat breeders at CIMMYT in Mexico attempt to obtain durable resistance by testing material at many locations throughout the world (Rajaram and Torres 1986). They assume that genotypes that are resistant everywhere have several genes for resistance and their resistance should be durable. Rajaram and Luig (1972) did find that resistance in lines that had low average coefficients of infection in international tests was complex in inheritance. The danger is that they may have one gene or a combination of genes for which a virulent genotype of the rust has not yet evolved. If the host genotypes are grown widely, virulent pathogen genotypes may well appear.

8.14.2 Breeding for Durable Resistance

Because of the way durability is defined, it is impossible to breed specifically for durable resistance. The best that can be done is to use parents that have been shown to have durable resistance and to breed for types of resistance that tend to be durable. Generally, this means breeding for polygenic resistance. However, resistance to stem rust controlled by the single gene *Sr26* has proven to be durable. Because of the difficulty in identifying durable resistance and the masking effect of genes for specific resistance that is not durable, few breeding programs appear to be breeding specifically for durable resistance. Nevertheless, it can be done.

All available information should be used to identify useful parents. If they do not carry adequate resistance, then they should be intercrossed, perhaps using multiple crosses, to obtain transgressive segregates with better resistance. Since genes for specific resistance can mask durable, polygenic resistance, a procedure must be used that eliminates their effects. Specific resistance is usually expressed in both seedlings and adult plants, while polygenic resistance is often expressed only in adult plants. This makes possible two procedures for eliminating the masking effects of genes for specific seedling resistance. If it is available, a rust isolate can be used that is virulent on all of the parents at the seedling stage. Then in the progeny of crosses involving the parents, selection for adult plant resistance in the field should select for durable polygenic resistance. If such an isolate is not available, then the early generation progeny of a cross can be tested for seedling resistance to the most virulent or most dangerous rust isolate available. All of the *resistant* plants must be eliminated. Since genes for specific resistance are usually dominant, one generation of selection may eliminate most of them. However, a second generation may be necessary, particularly for leaf rust for which more genes for resistance are recessive. The progeny of the susceptible seedlings should then be grown in the field and selected for adult plant resistance to the same race. Under either procedure, several generations of selection in the field may be necessary to obtain lines with good resistance. Resistance selected in this way will probably be polygenic but it could also be due to a gene or genes for specific adult plant resistance.

Although it is sometimes thought that durable resistance can only be moderate or incomplete, there is no reason why this must be the case. Krupinsky and Sharp (1979) intercrossed commercial wheat cultivars with varying levels of resistance. Even from crosses that had no resistant plants in the F_2 or F_3 generations, they were able to obtain plants with good levels of resistance after several more generations of selection.

Selection under controlled conditions for one or more of the components of resistance, after eliminating genes for specific, seedling resistance, may also be effective. Selection for a long latent period will probably be most effective. A long latent period can result in substantial reduction in a rust epidemic. It is particularly effective in areas where the rust season is very short and the rust must build up very quickly to cause serious damage.

8.15 Mutation Breeding for Rust Resistance

In early work on mutation breeding for rust resistance in wheat, there were some very optimistic reports of up to 2–3% of M_2 families (i.e., the second generation after treatment with a mutagen) carrying resistant mutants. However, in an experiment with oats, Caldecott et al. (1959) showed that the occurrence of supposed mutants for resistance to stem rust depended on whether the M_1 generation was grown near other oats or in isolation. When the M_1 was grown under isolation, no mutants occurred. In reality the reported mutants for rust resistance were mostly the products of mutagen-facilitated outcrossing. The treatments induced chromosome aberrations, aneuploidy and male sterility. When self-pollination failed in M_1 plants, florets opened and stayed open, and outcrossing occurred. Unfortunately, papers are still being published which give no indication that the authors recognize the need to begin with a highly purified seed source and to prevent any possibility of outcrossing, at least in the M_1 generation and preferably in later generations as well. Scientists who take such precautions rarely obtain rust-resistant mutants.

Specific resistance to the rusts is commonly controlled by dominant genes, indicating that resistance depends on a physiologically active gene product. The probability must be very small that a random mutation process will produce a change in a recessive gene that results in it producing an active gene product instead of an inactive one. The probability may be higher of producing mutations for resistance controlled by recessive polygenes. Unfortunately, the effects of such changes may be too small to be readily identified.

Mutagens do produce mutations, most of them recessive. The probability of producing genes for rust resistance that are not already available is small. If a situation arises where no other source of resistance is available, a mutation breeding program is worth trying. It should be carried out on a large scale and with the same resources as are required for any successful breeding program.

8.16 Biotechnology and Breeding for Rust Resistance

Rapid advances are being made in the field of biotechnology and some of them may have applications in breeding for disease resistance.

The growth of tissues, cells or protoplasts in culture and the regeneration of plants results in a large increase in variability – referred to as somaclonal variation (Larkin and Scowcroft 1981), or as protoclonal variation when plants are derived from protoplasts. In a number of species the regenerants include plants with disease resistance. Somaclonal variation has been reported in wheat (e.g., Larkin et al. 1984). Different genotypes differ greatly in culturability and frequency of regeneration.

Maddock and Semple (1986) obtained three lines of Brigand winter wheat which showed lower levels of leaf rust infection in the field but no seedling resistance. Their results suggest that the frequency of rust resistance will be low. In some host-pathogen systems it has been possible to use toxins from the pathogen to select for resistant variants in cultures. Such a procedure has not been developed for rusts. Somaclonal variation for rust resistance will be of interest only if the resistance is of a type not already available or if the procedure can be used to add resistance quickly to a desirable but susceptible cultivar. Otherwise it has no advantage over a method such as backcrossing.

Anther or pollen culture is another procedure which is used to speed up plant breeding. Cultures from F_1 plants from crosses are cultured to produce calli, the calli are induced to produce haploid plants which are then doubled using colchicine. There are several difficulties with anther culture in wheat. The frequency of regeneration is low and many of the resulting plants are albinos. Both problems vary considerably for different genotypes and are being gradually overcome by improvements in culture media. Although superior doubled haploids can be obtained (e.g., de Buyser et al. 1981), Baenziger et al. (1983) found that doubled haploids from the cultivar Kitt on the average yielded significantly less (16%) than the parent, a problem that has also been reported in tobacco. Despite the difficulties, anther culture has potential as a way of rapidly producing doubled haploids from a cross, increasing them and selecting for rust resistance and other desirable characters.

Potentially the most exciting of the new biotechnological procedures is the transformation of plant cells and protoplasts. For some species, procedures are already available for transferring a specific gene into a plant. Wheat and the monocotyledonous plants in general have proven to be more difficult to work with. Nevertheless, there is little doubt that efficient procedures will be developed. The ideal situation will be when wheat breeders can transfer genes for resistance at will into cultivars. Unfortunately, a great deal of work must be done before this will be possible. As yet no gene for rust resistance in wheat has been isolated and cloned, a first essential step in transformation. More information is needed on the biochemical pathways that result in rust resistance.

8.17 Hybrid Wheat

Despite the large amount of research that has gone into the production of hybrid wheat, the area planted with hybrids is still relatively small. Work is continuing with systems using cytoplasmic male sterility and there is increasing research on the use of chemical hybridizing agents.

Hybrids must meet the same requirements for resistance to rusts and other diseases that any cultivar does. Since many major genes for specific resistance are dominant, it is easier to combine a number of genes in a hybrid than in a homozygous cultivar. If each parent carries several different dominant genes for resistance, then all of the genes should be expressed. Furthermore, in a hybrid it is possible to combine two alleles at one locus. However, if it is particularly important to have recessive genes expressed in a hybrid, then they will have to be present in both parents. This is particularly likely to be the case for types of resistance that are controlled by several genes each having a small effect, types that may be durable.

In developing possible parents for producing hybrids, careful planning is necessary to ensure that the required disease resistance is present. Because of the situation that developed in maize, in which all hybrids carrying Texas male-sterile cytoplasm proved to be susceptible to race T of southern corn leaf blight (*Helminthosporium maydis* Nisik and Miyaki), it is desirable to use more than one type of male-sterile cytoplasm in developing hybrids. Washington and Maan (1974) found that cytoplasms from different *Triticum* species changed the reaction to leaf rust of some genotypes. One advantage of the use of a chemical hybridizing agent is that only normal wheat cytoplasms are involved.

8.18 Selecting a Method of Breeding for Resistance

Deciding on the method or methods to use in breeding for rust resistance in a particular area is not simple. The decision is affected by the importance of each rust in the area; their life cycles, particularly how they survive the critical period of the year (over winter or over summer) and the number of urediospore cycles that occur; the parental material that is available; and previous experience with resistant cultivars. In areas where the rusts are not endemic and epidemics occur only occasionally, simple breeding procedures using specific resistance may be adequate. This has largely been true for the control of stem and leaf rust in North America. The rusts are much more difficult to control in subtropical areas where rust is endemic and can go through many urediospore cycles on a single crop, particularly if it is grown over the winter. Then it may be essential to use more genetically complex resistance, either combinations of genes for specific resistance or polygenically controlled durable resistance. In addition it may be necessary to manage the use of genetic diversity as described in Chapter 9.

CHAPTER 9

Managing Genetic Diversity to Control Rusts

9.1 Introduction

Until fairly recently, in many countries cereal crops consisted largely of landraces that were often mixtures of diverse genotypes. Even mixtures of crops such as wheat, oats, and barley were fairly common, and are still grown in some areas. This genetic diversity often provided protection against diseases. The more recent emphasis on a few, high-yielding, very uniform cultivars in crops has greatly increased their genetic vulnerability to disease. In natural ecosystems there is both interspecific and intraspecific diversity. The pathogenicity of a pathogen and the reaction of a host evolve in such a way that they coexist. The host does not become so resistant that it wipes out the pathogen and the pathogen does not become so virulent and aggressive that it wipes out the host. In modern cereal production large areas are planted to single, genetically very uniform genotypes, often for a number of years. Plant breeders and farmers control the "evolution" of the host and force the evolution of pathogens in specific directions, sometimes with disastrous effects.

Genetic diversity can be employed at three levels, intrafield, interfield, and interregion, to manage the evolution of a pathogen. The objective is not to eliminate the pathogen but to produce an equilibrium between host and pathogen that results in little damage to the host and little change in the pathogen.

Groenewegen and Zadoks (1979) pointed out that in agricultural crops of mixed genotypes there is a continuum of possible levels of diversity from mixtures of species to mixtures of isogenic lines. They describe the characteristics of five types of mixtures, particularly in relation to disease control.

1. Mixtures of species show the maximum amount of variability. They have diversified resistance and provide broad and durable resistance to many diseases.

2. Mixtures of cultivars may show considerable variability, depending on the similarity of the cultivars. They may provide resistance to several diseases, depending on the resistances of the cultivars that are selected.

3. Mixtures of moderately related lines with 50–90% of their genes in common will show moderate variability, depending in part on the amount of selection that is practiced. The lines are usually selected for diverse resistance to one target disease but resistance to other diseases is possible, depending on choice of parents and selection in crosses.

4. Mixtures of closely related lines produced by 3–7 backcrosses and having 90–99% of their genes in common will be quite uniform. The lines are normally selected for diverse resistance to only one disease.

5. Mixtures of isogenic lines produced by eight or more backcrosses will have almost 100% of their genes in common except for the resistance genes and genes

linked to them. They should be very uniform except for diverse resistance to the target disease, but provide an unnecessary degree of genetic uniformity.

Types 2, 3, and 4 are currently being used to control rusts in wheat. Both 3 and 4 are often included as types of multilines differing only in the amount of backcrossing used in their development (see Sect. 9.2).

Mixtures of resistant and susceptible genotypes can slow the development of epidemics in several ways:

1. In a mixture, susceptible plants are farther apart and fewer spores will spread from one to another.

2. Resistant plants can act as a physical barrier to spore dispersal, particularly in adult plants when the canopy is heavy. Many spores will move from a susceptible plant to a resistant plant which they cannot infect (often called spore trapping).

3. Infection with an avirulent genotype may induce resistance to a virulent genotype. Several studies have reported the induction of resistance to rust in cereals including wheat. Cheung and Barber (1972) found that inoculation with an avirulent race of stem rust followed 3 to 6 days later by a virulent race reduced the number of pustules by about 80%. Johnson and Taylor (1976) showed with yellow rust that simultaneous inoculation on the same side or on opposite sides of the leaf resulted in a reduction in spore production of up to 56.5%. The effect varied with different cultivars. If inoculation with the virulent race was delayed for 4 days, the reduction in spore production was from 51.4 to 65.2%. Johnson (1978) was able to demonstrate small effects with yellow rust in the field.

9.2 Multilines

In discussing oat breeding, Jensen (1952) emphasized the danger of genetic uniformity for reaction to oat diseases and recommended diversification. Because of the requirement for uniformity in seed certification programs, he proposed the development of a multiline cultivar composed of pure lines that had been selected for uniformity of appearance but differed in genotype. Jensen suggested that a multiline would provide greater protection against disease, although he was thinking in terms of resistance against several diseases rather than against several races of one disease.

Borlaug (1959) carried the procedure a step further by proposing the use of backcrossing to develop uniform lines carrying different genes for resistance to stem rust. Eight to 16 such lines would then be combined to produce a multiline cultivar. A program was started at CIMMYT using standard height cultivars as the recurrent parents. However, the material was discarded when the development of semidwarf cultivars made it obsolete. This experience illustrates one of the dangers in developing multilines. The procedure is lengthy and by the time the lines are developed the recurrent parent may have been replaced by a better cultivar.

Some confusion has arisen over the use of the terms multiline and mixture. Jensen (1952) used the term multiline to refer to a mixture of lines that had been selected for uniformity of appearance but were otherwise quite different genetically. Since then multiline has come to be used most frequently for mixtures of lines that are fairly similar genetically and have often been produced by backcrossing. It will be used in that sense in this book. Mixtures of cultivars will be called mixtures. Obviously, there is no clear distinction. Depending on how components are produced they may range from being near-isogenic to being largely unrelated genetically.

9.2.1 Clean and Dirty Multilines

Borlaug's (1959) original idea was to produce a series of backcross lines that were resistant to all of the rust races prevalent in an area. A multiline made up of such lines would be free of rust or "clean." Additional lines would be produced and kept in reserve. If during commercial production one component became susceptible, it would be replaced and a new version of the multiline released.

A major problem in producing a clean multiline is obtaining enough genes or different combinations of genes that provide resistance to all the prevalent races in an area. Fortunately, it was soon realized that it was not necessary to have a multiline that was completely free of rust. A moderate amount of rust, particularly late in the season, would have little effect on a crop. For example, suppose that on a susceptible crop an initial spore shower results in two pustules for every ten plants and that the spores from each pustule produce 50 new pustules. In the second cycle there would be ten pustules per plant and in the third 500 per plant, assuming that there is no overlapping of cycles and spores disperse randomly. However, if the same spore shower landed on a multiline and only 50% of the plants were susceptible, in the first cycle there would be only one pustule for every ten plants. This would build up to 2.5 pustules per plant in the second cycle and 62.5 per plant in the third. Thus, after three cycles the epidemic on the multiline would be only one-eighth as heavy as that on a susceptible crop and would probably cause little damage. A multiline in which the components are susceptible to some races is called a "dirty" multiline. For every prevalent race, some components of a dirty multiline must be resistant, otherwise a race virulent on all the lines could cause damage. Some components may be resistant to all races.

9.2.2 Producing Lines for a Multiline

The choice of the recurrent parent on which to base a multiline is very important. It must be the best cultivar available and one that is likely to last for some time. If the best cultivar has serious faults and will obviously be replaced quickly, there may be little point in starting a multiline program based on it. Fortunately, there is often a good cultivar available whose main fault is rust susceptibility.

In choosing donor parents, the key is to obtain as many different genes for rust resistance as possible. Usually simply inherited, specific resistance is used, since complex resistance is difficult to follow in backcrosses. Donors should be selected from material of diverse parentage that has been tested widely. Testing them against as many races as possible can provide valuable information about the genes they carry and whether they carry the same genes. Genetic studies are, of course, very valuable but are not often available. How well adapted the donors are is of little importance except that poorly adapted donors may require more backcrossing.

The number of backcrossess required to produce the lines varies greatly depending on the circumstances. If uniformity is critical or very specific quality requirements must be met, then a number of backcrosses are essential, perhaps as many as five or more. In countries that have plant patent legislation a number of backcrosses will be necessary if multilines are to meet the requirements for uniformity and stability. If great uniformity in agronomic or quality characteristics is not essential, then one or two backcrosses may be sufficient. The use of limited backcrossing makes it easier to select the material for more than just rust resistance. In India, several programs have been carried out using Kalyansona as the base parent, often with only one or even no backcrosses. For example, Gill et al. (1980) produced a large number of lines related to Kalyansona and selected six to combine in a multiline. The components were similar to Kalyansona in height and maturity, but had heavier seeds and outyielded it by about 13%. They were generally resistant to both yellow rust and leaf rust.

During the backcrossing it is essential to test the progeny with rust races that will allow selection for the resistance carried by the donors. The recurrent parent must, of course, be susceptible to them. Seedling tests with single races can often be used. Field tests with natural epidemics may be unreliable in detecting specific genes for resistance because of the uncertainty about the races that are present.

When homozygous resistant lines have been obtained, they should be evaluated for both rust resistance and agronomic characters. The fewer the backcrosses the more testing that should be done in order to obtain well-adapted lines that are relatively similar in appearance. Occasionally a gene for resistance or genes linked to it will have deleterious effects and a line will not be equal to the recurrent parent. The reverse may also occasionally be true and a line will be better than the recurrent parent. Rust tests with single races and natural field epidemics are desirable in order to determine the usefulness of the lines.

Finally, a number of desirable lines should be selected to combine in a multiline. There is no ideal number but between 6 and 12 seem to be commonly used. The number should not be too low because then one line makes up a large percent of the population and this could be a problem if it became susceptible. However, a smaller number makes it easier to increase the lines and bulk them to produce the multiline.

9.2.3 How Many Lines Must be Resistant to Each Race?

In formulating a dirty multiline, a key question is, "What percentage of the population must be resistant to any race in order to provide adequate resistance?" Data from Israel showed that in natural ecosystems the most virulent race group of oat crown rust (264–276) averaged 40% of collections while about 30% of the *Avena sterilis* population was resistant (Browning et al. 1979). Apparently the host and pathogen populations reached an equilibrium at these levels. In the Iowa program to produce crown rust-resistant oat multilines, the goal is to have 60% resistance to each race (Frey et al. 1979). This is accomplished by varying the amount of each component in the multiline. For example, Multiline E74 was composed of 4% of each of five lines, 18% of each of two lines, and 22% of each of two lines. Twenty-two percent of its components were susceptible and 26% of moderately susceptible to biotype 326. Other evidence has suggested that having 30–40% of the plants resistant to a particular race will be sufficient. This may vary somewhat depending on the prevalence of a race in the initial inoculum and on how favorable the environment is for the rust.

9.2.4 Production of Seed of Multilines

Seed production of a multiline involves two steps: first, increasing the seed of the lines separately, and second, mixing the lines and increasing the mixture. The more generations of the multiline that are grown, the greater the danger of changes in the frequencies of the components. For this reason it is desirable to produce as large increases of the individual lines as is practical. If the lines have similar weights per seed and germination, equal weights of each lines can be mixed. Otherwise, corrections for weight per seed and germination should be made so that equal numbers of viable seeds of each line are mixed.

If the lines have all been backcrossed several times to a recurrent parent, they should be relatively similar and interline (intergenotypic) competition should be minimal. Thus, it should be possible to increase the multiline for several generations without greatly changing its composition. Several generations of increase may be necessary. Suppose for example that one tonne (1000 kg) of each of ten lines is produced and mixed to give 10,000 kg of seed of the multiline. At 100 kg/ha, this will seed 100 ha. At a yield of 3 tonnes/ha (a 30-fold increase), in two generations there will be enough seed for 90,000 ha. Depending on the yields, seed production could be considerably more or less than this.

9.2.5 Changes in the Composition of a Multiline

In many areas, farmers will want to use their own seed for several generations. The amount of change in the composition of the multiline should depend on the similarity of the lines and the amount of intergenotypic composition. Murphy et al. (1982) mixed equal quantities of five crown rust-resistant oat lines. Four had been backcrossed five time and one four times to a recurrent parent. Over five generations, sizeable changes occurred in the composition of the multiline (Table 9.1). The changes were greater in a rust-free environment (sprayed with Maneb) than when rust was present. However, the trends were the same in both environments. CI 9184 showed the largest increase in each, although the regression coefficient showing the change in percent per generation was not significant in the rust-free environment because of large deviations from the regression line. CI 9192 showed the largest decreases in each environment. Murphy et al. (1982) stated that it was "recognizably inferior" and probably would not have been included in a normal multiline. Considering the number of backcrosses that had been made and the resulting genetic similarity of the lines, the changes are surprisingly large.

In a normal field situation, selection could favor the most resistant or highest-yielding components of a multiline and actually improve its performance. Alternatively, it is possible that a multiline could stand considerable change in its

Table 9.1. Changes in the composition of an oat multiline over five generations. (Murphy et al. 1982)

Environment	Near-isogenic line (NIL)	Percent of NIL in generation 0	1	2	3	4	Regression coefficient[a]
Rust-free	CI 9192	22	18	16	14	10	-2.80**
	CI 9183	18	23	19	19	14	-1.20
	CI 9184	20	22	31	26	38	4.00
	CI 9190	21	23	21	24	20	-0.10
	CI 9191	19	14	13	17	18	0.10
Rusted	CI 9192	22	19	19	14	15	-1.90*
	CI 9183	18	20	19	17	22	0.50
	CI 9184	20	22	25	28	28	2.20**
	CI 9190	21	23	25	24	19	-0.30
	CI 9191	19	16	12	17	16	-0.50

*,**Significant at the 0.05 or 0.01 levels of probability, respectively.
[a]The regression coefficient showing the change in percent per generation.

composition without serious effect on its rust resistance or yield. More studies to clarify the situation would be very helpful.

9.2.6 The Effect of Multilines on Rust Epidemics

Vanderplank (1963) proposed that the early stages of development of a polycyclic or "compound interest" disease such as rust can be described by the equation

$$X_t = X_0 e^{rt},$$

where X_t is the disease severity at time t, X_0 is the initial severity, r is the apparent infection rate, and t is the time since the initiation of the epidemic. The equation is appropriate for the period when growth is essentially exponential. However, as the severity of the disease increases, there is less uninfected tissue in which new infections can occur and r declines.

A major effect of multilines is to reduce X_0 compared to the value for a completely susceptible population. In a susceptible population, all of the initial inoculum will land on susceptible plants and may cause infection. In a multiline in which 50% of the plants are susceptible, only half of the spores will land on susceptible plants and the remaining half will be lost. Thus, X_0 will be reduced by one half. This reduction in the initial severity can be particularly important if the period of disease development is short so that the pathogen goes through relatively few cycles of reproduction. The spores that are produced on the components in a multiline will also go through the same spore-trapping process. Some will land on resistant plants and fail to produce infections, thus reducing the apparent infection rate. However, because of the frequency with which spores from a plant reinfect it, the reduction in r may not be as great as might be expected. Nevertheless, the combined effects of reductions in X_0 and r can be substantial.

Luthra and Rao (1979) tested six multilines containing from 4 to 12 leaf rust-resistant components based on Kalyansona. The multilines and the compo-

Table 9.2. The severity of leaf rust infection and yield of six multilines based on Kalyansona. (Data from Luthra and Rao 1979)

Multiline or parent	Number of components	Proportion of susceptibility	Final rust severity	Yield-tons per hectare
MLKS-7506	7	0.036	5.80	4.47
MLKS-7501	4	0.063	8.79	4.75
MLKS-7505	7	0.068	7.80	4.47
MLKS-7507	12	0.139	0.95[a]	4.82*
MLKS-7502	5	0.183	28.91	4.39
MLKS-7503	5	0.304	35.00	4.46
Kalyansona	1	0.833	–[b]	4.24

[a]Apparently a typographical error since the reading in the preceding week was 5.99 (possibly 10.95%).
[b]Not given but must have been at least 70% based on the average of the components for MLKS-7502, in which it is one component.
*Significantly higher than Kalyansona.

nents were compared in a six-replicate yield test grown in a nursery inoculated with 12 leaf rust races. Table 9.2 is derived from their data. The proportion of susceptibility was calculated by multiplying the proportion of each component in the multiline by the proportion of races it was susceptible to, e.g., for MLKS 7501, one of the four components was susceptible to two races and one to one race – $(0.25)(2/12)+(0.25)(1/12)=0.063$. All of the multilines showed substantially reduced rust severity compared to Kalyansona and the severities correlated well with the proportion of susceptibility ($r=0.94$*). The rust epidemic developed late and yield losses were small. All of the multilines gave slightly higher yields than Kalyansona but only one was significantly higher.

9.2.7 Experience with Multilines

The first successful multiline breeding project in wheat was carried out by the Rockefeller Foundation in Colombia (reviewed by Rajaram and Torres 1986). Over 600 donors of resistance to yellow rust and stem rust were backcrossed three times to the Brazilian cultivar Frocor. A large number of lines were produced and tested and the best ten were combined into Miramar 63. It apparently remained resistant to both rusts for a number of years but usage declined because of problems with seed multiplication and certification.

The most extensive use of multilines in cereals has been for resistance to crown rust (*Puccinia coronata* Corda) in oats in Iowa. Despite many years of breeding for resistance, crown rust was causing annual losses of 18–30% in grain yield (Frey et al. 1979). Two series of multilines were produced, one based on an early parent and one on a mid-season parent. By 1979, 13 different multiline mixtures had been released (Browning et al. 1979). They were grown on substantial areas and remained resistant. The rust season is relatively short in Iowa (30–40 days) and a delay in epiphytotic development is sufficient to provide control. An experiment in Texas, where the season is 4 months long, showed that the multilines provided control there as well (Browning et al. 1979). However, the area planted to the multilines declined as agronomically superior oat cultivars were developed, crown rust pressure declined, and the barley yellow dwarf virus (BYDV) became important (Browning and Frey 1981). However, new multilines based on the BYDV-resistant cultivar Lang were produced.

Tumult, a winter wheat multiline with resistance to stripe rust, was released in The Netherlands in 1975 (Groenewegen 1977). Yellow rust resistance was transferred to Tadorna from seven sources and five of the resulting lines were used to produce Tumult. Two lines were kept in reserve.

Allan et al. (1983) produced a single-gene, semidwarf multiline, Crew, based on semidwarf derivatives of Omar winter club wheat. It is made up of ten components, five resistant, two moderately resistant, and three intermediate in resistance to yellow rust. A minimum of nine genes is probably involved. Three components have some resistance to leaf rust and four to powdery mildew. In a comparison between fungicide-treated and untreated plots in three tests, Crew showed a maximum loss of 7% compared to 9 to 27% for its components.

As described by Rajaram and Torres (1986), the wheat breeders at CIMMYT in Mexico are using a modified multiline system to breed for resistance to several diseases. The procedure is based on derivatives of 8156 such as Siete Cerros, and involves at most limited backcrossing. Segregating populations are subjected to intense selection for disease resistance. High yielding lines are grouped into clusters having different disease resistance, and are classified for maturity, height, grain color, etc. Sets of six to ten lines that are similar in many characters but differ in disease resistance are then put together and tested.

9.2.8 The Effect of Multilines on the Spectrum of Rust Genotypes

The development and use of multilines posed some interesting questions about their potential effects on the frequencies of rust genotypes. Some scientists argued that the use of a number of genes for resistance in a single multiline would force the evolution of a superrace capable of attacking all the genes. This would be particularly likely to happen if individual lines carried single genes that could be overcome one at a time by mutation in the pathogen. Other people argued that most genes for virulence are recessive, recessive genes are often deleterious and, therefore, stabilizing selection as proposed by Vanderplank (1963) would favor simple races with few genes for virulence. A number of studies have shown that complex races are not necessarily less fit, and this is not surprising. Even if genes for virulence tend to be deleterious they are almost certainly only a few of the many genes that affect the fitness of a genotype. Over time, the fitness of a particular rust race may change as selection favors background genotypes that are more competitive.

Several authors have developed simulation models to attempt to predict the likelihood of a superrace developing. As with all such models, the predictions are only as good as the assumptions that are made. The most important assumption is probably the size of the selection coefficients for different rust genotypes in the absence of rust. Unfortunately, it is very difficult to measure selection coefficients for different genotypes under field conditions. For one thing, little is known about the alleles carried by rust isolates at many loci for pathogenicity, let alone other genes affecting competitiveness that they may carry. In some cases very different selection coefficients have been calculated for the same gene in different experiments.

Marshall et al. (1986) did extensive computer simulation. Examination of their tables shows that selection coefficients against genes for virulence must be fairly large (≥ 0.20) to prevent the development of complex races. In a field population of a rust it seems unlikely that genes that have a sizeable selective disadvantage in the absence of rust would persist unless rust occurred regularly. However, some unnecessary genes for virulence do persist in populations, indicating that they do not have a selective disadvantage. This suggests that the multilines could tend to force the evolution of complex races. Nevertheless, the effect of dirty multilines on the composition of rust populations is still an open question. It may be answered only by the long-term cultivation of multilines on a large scale. The limited experience so far does not indicate that superraces are produced.

9.2.9 Problems with Multilines

Multilines have two major problems. First, their production takes a considerable amount of time and effort. By the time a multiline is ready for release, the recurrent parent and, therefore, the multiline, may be obsolete. Second, in most cases a multiline is developed for resistance to only one disease and in many cases several diseases are important. Both of these problems can be at least partially overcome if great uniformity and, therefore, extensive backcrossing is not required. If the number of backcrosses can be reduced, then the time to produce a multiline can be reduced. With limited backcrossing, it is easier to select for resistance to more than one pathogen, as is done in the CIMMYT program (Rajaram and Torres 1986).

If multilines are developed and grown widely, and are found to force the evolution of superraces of the rusts, this will be a third and very serious problem. By the time it is apparent that superraces are evolving, considerable damage may have been done. Controlling the superraces could be difficult.

9.2.10 Where to Use Multilines

If multilines are successful, their major effect should be to stabilize rust populations. They should be most valuable in areas where the rapid evolution of new races is a major problem. Thus, they may be of particular value in areas where a rust is endemic, or in overwintering or oversummering areas. Stabilization of the rust population in these areas should make the job of the wheat breeder much easier in surrounding areas.

9.3 Cultivar Mixtures or Blends

Mixtures of cultivars are intended to serve the same purpose as multilines – the use of genetic diversity to control plant diseases. They have the advantage that they utilize existing cultivars so that no special breeding is necessary and the components can be changed as new cultivars become available or the disease situation changes. In addition, they maximize the amount of genetic diversity within fields of a single crop. With multilines it is usual to have one target disease, although it is possible but more difficult to breed for resistance to more than one. With mixtures a broad range of cultivars differing in resistance to several diseases can be considered as possible components. Because of this diversity, combinations can be selected that provide resistance to two or more diseases. Furthermore, it may be possible to select cultivars that complement one another and thus increase yield.

So far cultivar mixtures have been used mainly in barley, primarily to control powdery mildew, and to a lesser extent in wheat to control yellow rust and powdery mildew (Wolfe 1978, 1985). They have involved three-cultivar mixtures. Wolfe et al. (1981) found that the mildew severity (%) in four barley mixtures averaged 28% of that of the components grown separately (range 3–43%). They also showed that the resistance was due not only to identified resistance genes but also to unidentified resistances. Although first tested commercially in England, the largest area of barley mixtures is now being grown in East Germany as a way of reducing the need for and cost of fungicides (Johnson and Lupton 1987).

The value of mixtures to control rusts in wheat will depend on the importance of uniformity in commercial fields and in the end use of the grain. Under present systems wheat is rarely kept segregated by cultivar. Mixtures of cultivars, all of which are suitable for a specific end use, should be acceptable.

In some areas, the short time between the harvest of one winter wheat crop and the planting of the next may cause problems in seed handling. It may be difficult to harvest the cultivars, mix them and distribute the mixtures to farmers in time for seeding. However, it should be possible to reproduce the mixture for one year and use the resulting seed for the next crop. Because of the greater variability in a mixture than in a multiline, larger changes in its composition may occur as a result of competition among the components over several generations. More data are needed to determine whether this will be a problem if farmers use their own seed for a number of generations.

Fungicides can be used in conjunction with mixtures (Wolfe 1985). For example, seed of the most susceptible component could be treated with a systemic fungicide. This would enhance the resistance of the mixture and reduce the amount of fungicide used compared with treating all the seed.

As with multilines, the increasing, mixing, and distribution of seed may be a problem in areas where seed-handling facilities are not well developed. In East Germany it has been suggested that the component cultivars for a mixture be planted in alternate drill strips. At harvest, a combine can be driven back and forth across the strips, producing seed already mixed for planting.

Mixtures provide an interesting way of using genetic diversity to control wheat diseases. They do not require much work to produce and they can constantly change and evolve as new cultivars are produced or in response to changes in pathogens. Their use should be considered wherever uniformity is not a problem.

9.4 Interfield Diversity

To minimize the danger from rapid changes in pathogen populations, several countries recommend that farmers plant cultivars with different resistances in adjacent fields. For example, in England Priestley (1981) described a system in which wheat cultivars were classifield into diversification groups, each containing cultivars that were genetically similar for resistance to yellow rust. It was recommended that farmers plant cultivars from different groups in different fields. If their neighbors did the same, then the build-up of the pathogen would be slowed as it constantly met fields of different genotypes. Analysis showed that spores derived from a cultivar possessing one type of specific resistance were largely avirulent on a cultivar carrying another type of resistance. Thus, there was little danger of a sudden increase in one race as can happen if one cultivar predominates in an area.

Recently the system has been extended to provide diversity among fields for resistance to both yellow rust and powdery mildew (Johnson and Lupton 1987). Tables are published each year showing farmers which combinations of cultivars will restrict the spread of the two diseases (e.g., Table 9.3). Unfortunately, the scheme does not seem to be widely used by farmers in the United Kingdom (Johnson and Lupton 1987).

Table 9.3. Part of a table of 14 winter wheat cultivars showing which combinations in adjacent fields will reduce the spread of yellow rust and powdery mildew. (Johnson and Lupton 1987)

Cultivar	6	7	8	9	10
6. Norman	ym	+	+	+	m
7. Armada	+	ym	+	+	+
8. Slejpner	+	+	y	y	+
9. Stetson	+	+	y	ym	+
10. Galahad	m	+	+	+	ym

\+ A good combination that should reduce the spread of both yellow rust and powdery mildew.
y Risk of spreading yellow rust.
m Risk of spreading powdery mildew.

9.5 Regional Deployment of Genes for Resistance

In some wheat-producing areas of the world, one or more of the rusts is endemic and survives throughout the year. More commonly, however, the rust survives a critical season of the year in only limited areas and then spreads over large areas as conditions become favorable again. Often the critical season is winter in a cold climate and summer in a hot climate. Stem rust in North America is a prime example. In most years the rust overwinters only in southern Texas and possibly some southeastern states. As weather conditions become favorable in the spring and crops begin to grow, the rust increases and then is blown successively farther north until it runs out of wheat fields on the northern edge of the Canadian prairies. In total, the rust moves over 3000 km. In the fall, when winds are favorable, spores are carried back from the north to the overwintering area to start the cycle again.

The long Puccinia path in North America, and similar paths elsewhere, provide an opportunity to deploy resistance genes in a way that will slow the spread and delay the development of an epidemic. If, as it moves along a Puccinia path, the rust pathogen meets constantly changing genotypes for resistance in the host, then at any point a large proportion of the spores will land on resistant plants. Each new set of host genotypes with different genes for resistance will act as a selection sieve. Only pathotypes that are virulent will pass through the sieve. In North America, where the rust season is often short, particularly in the north, slowing the rust build-up by reducing the initial infection rate can be sufficient to prevent damage.

For such a system to be effective, the deployment of resistance genes in particular areas must be controlled by an agreement among the wheat breeders involved or by the assignment of genes to specific areas by an acceptable authority. That is not easy. Many breeders may be involved, often in more than one country. Frequently, the genes carried by the parents being used are unknown. Nevertheless, breeders in one area can improve the situation by deliberately looking for new sources of resistance and sources that are not being used by other breeders along their Puccinia path.

Oat breeders set up a regional deployment scheme for oat crown rust in the great plains of North America, an area in which there are major oat breeding projects in Texas and Iowa in the United States, and Manitoba in Canada. Initially, three genes were assigned to each of the northern and central regions and four to the southern (Frey et al. 1973). No known crown rust race was virulent on resistance genes in more than one area. However, because of a decline in the area planted to oats and in the pressure from crown rust, the scheme was never implemented (Mundt and Browning 1985).

No similar scheme to control either leaf or stem rust in wheat has been established in North America. Nevertheless, there is some unintentional gene deployment. In general, the gene pools in spring and winter tend to be fairly distinct. Some genes for rust resistance are used in one type and not in the other. For example, *Sr6* has been widely used in spring wheat but not as widely in winter wheat. One race of stem rust has predominated in North America in recent years. Nevertheless, some marked changes in the virulence of the stem rust population occurs as it makes its annual migration from south to north. For example, in 1983 virulence on several genes for resistance increased from 65–75% in the south to 99–100% in the north while virulence on two others declined from 34 and 30% to 1%. The changes were almost entirely due to the presence of race 151-QCB in the south but not in the north (Roelfs et al. 1984).

The overwintering area is critical in the North American Puccinia path. On the one hand, only races that are virulent on the cultivars grown in the area will survive. On the other hand, any resistance gene that is used extensively in the overwintering area will almost certainly be overcome by a virulent genotype of the pathogen. As a result, it will be of no use in the rest of the Puccinia path. It could be argued that it would be economically viable to subsidize farmers in the overwintering area to grow very susceptible cultivars on which, hopefully, only simple races with few genes for virulence would survive. These races would then be unable to attack the more resistant cultivars farther north. The absence of rust farther north might then reduce the chances of rust being blown back into the overwintering area and becoming reestablished in the fall.

In the absence of such a strategy, extreme care should be used in selecting resistance to be used in the southern United States or in any area in a Puccinia path where rust survives a critical period. The aim should be to use resistance that stabilizes the rust population. Multilines are a possibility, if in fact they stabilize the rust population without forcing the development of a superrace. That, however, is still not known. A safer possibility is to use some form of resistance that is not controlled by major genes for specific resistance but is polygenic and effective against all races.

In India, leaf rust has become a major problem because of the breakdown of resistance in many semidwarf cultivars. Reddy and Rao (1979) proposed a system for deploying resistance genes and gene combinations in three areas of the country.

9.6 Conclusions

A variety of methods of using genetic diversity to control the rusts are now available to plant breeders. They are likely to be of most value in those areas where the rusts have been particularly difficult to control. Which method or combination of methods to use will vary, depending on the circumstances in a specific area. Certainly more consideration should be given to their use as a way of making better use of genetic resources and providing more reliable control of the rusts.

CHAPTER 10

The Transfer of Rust Resistance from Alien Species to Wheat

10.1 Introduction

The cultivated wheats belong to the tribe Triticeae of the grass family, the Gramineae. The basic chromosome number in the Triticeae is seven, suggesting that the species are all derived from one basic ancestor in the distant past. However, as a result of a long period of evolution, the Triticeae is now a very diverse complex. It can be divided into subtribes, two of which are the Triticineae which includes *Triticum* and *Secale* (rye), and the Hordeinae which includes *Hordeum* (barley) and *Elymus* (wild rye).

Wide crosses and polyploidy have played a major role in the evolution of the Triticeae. It is not surprising, therefore, that many interspecific and intergeneric crosses among the Triticeae have been successful (Knobloch 1968; Sharma and Gill 1983).

10.2 Relationships Among Species

The complexity of the Triticeae with many polyploids of different types, and the extensive hybridization among species, has led to taxonomic confusion. Löve (1982) and Dewey (1984) have proposed a genomic classification of the Triticeae, which is particularly useful for intergeneric hybridization. Under the genomic system, only species with the same genomic constitution are included in a genus. The system tends to make classification much less subjective than a system based on morphology. However, it also makes the construction of taxonomic keys based on morphology more difficult.

Sears (1981) grouped the relatives of bread wheat according to the closeness of their relationship to it (Table 10.1). The groupings provide a useful basis for determining the probability of success in interspecific and intergeneric transfers.

10.3 The Relatives of Wheat as Sources of Rust Resistance

The wild relatives of wheat contain a wide range of disease resistance, including resistance to the rusts. *Triticum timopheevii* is particularly noted for disease resistance but rust resistance has also been identified in many other relatives of wheat. For example, Kerber and Dyck (1979) tested 85 accessions of *Aegilops squarrosa* (*T. tauschii*) in field tests and found that 10 had resistance to stem rust and

Table 10.1. The relatives of wheat grouped according to the genomes they carry and the presumed closeness of their relationship to bread wheat. (After Sears 1981; Kimber and Sears 1987)

Group	Species	Genome formula
1. Species carrying only the A, B, or D genomes.		
(a) The diploid progenitors	*T. monococcum*	A
	T. tauschii	D
(b) The tetraploid progenitor	*T. turgidum*	AB
2. Polyploids with one homologous genome		
(a) The A genome	*T. timopheevii*	AG
(b) The D genome	*T. cylindricum*	CD
	T. ventricosum	DUn
	T. crassum	DM or DDM
	T. syriacum	DMS
	T. juvenale	DMU
3. Species with only homoeologous genomes		
(a) Closely related species	*T. speltoides*	S
	T. bicorne	S^b
	T. longissimum	S^l
	T. searsii	S^s
	T. kotschyi	US
(b) Less closely related species	*T. dichasians*	C
	T. comosum	M
	T. tripsacoides	Mt
	T. uniaristatum	Un
	T. umbellulatum	U
	Other U-containing polyploids (see Table 1.2)	
	Several *Elytrigia* species	
(c) Distantly related species	Species of *Secale, Haynaldia, Hordeum, Agropyron, Elytrigia,* etc.)	

38 to leaf rust. Vallega (1979) tested up to 102 accessions of *T. monococcum* with two or more races of each rust. All accessions tested were resistant to five races of leaf rust and many had resistance to yellow rust or stem rust. Grama et al. (1983) tested 750 accessions of *T. dicoccoides* (*T. turgidum*) and found that many contained plants that were resistant to yellow rust. Gill et al. (1983) tested large numbers of accessions of four species for seedling resistance to one isolate of leaf rust. The frequency of resistance was 225 of 547 accessions of *T. boeoticum* (*T. monococcum*), 106 of 241 of *T. araraticum* (*T. timopheevii*), 1 of 224 of *T. dicoccoides* and 68 of 198 of *T. urartu* (*T. monococcum*). Knott (unpublished) tested a collection of 185 accessions of 22 *Triticum* species to single isolates of leaf rust and stem rust. One hundred and thirteen were resistant to leaf rust and 103 to stem rust. Several accessions were susceptible to leaf rust as seedlings but resistant as adult plants. This sampling of results illustrates the tremendous potential of the relatives of wheat as sources of rust resistance.

Since rust resistance is readily available in the wild relatives of wheat, the next question is, "How widely can they or should they be used?" Even the simplest transfer of a gene from a close relative to wheat will be more difficult than a transfer from a resistant cultivar. For this reason, a wheat breeder who is looking for resistance should first search among available cultivars, land races, etc. If good resistance is difficult to find, then the search should be broadened to include the closest relatives of wheat. Scientists whose primary objective is not the immediate production of cultivars can consider the transfer of genes from alien species in order to broaden the gene pool available to breeders. In the process, valuable information can be obtained about the origin, evolution, and genetic structure of wheat.

When some of the early work on the transfer of rust resistance was started, particularly using some of the more distant relatives of wheat such as *Agropyrons* (*Elytrigias* and *Thinopyrums*), it was hoped that the resistance might prove to be durable. However, there is little evidence that resistance in the wild relatives of wheat is genetically or physiologically different from that in cultivated wheat. Consequently, there is no reason to think that it will be more durable, and in a number of cases resistance from an alien species has been overcome by a new, virulent biotype of a rust. However, resistance from an alien source is often initially effective against a wide range of rust races. This may be because the resistance had previously been present in only a small population of plants that had little contact with the rust. Thus, there had been little opportunity for virulence to evolve, but it may evolve if cultivars carrying the resistance become widely grown. Nevertheless, one gene for resistance to stem rust, *Sr26* from *Agropyron elongatum* (*Thinopyrum elongatum*), has been used extensively in Australia and so far has remained resistant. Luig (1983) reported that several attempts using mutagens to produce virulent mutations in the rust were unsuccessful. However, later (personal communication) he indicated that a culture virulent on *Sr26* had been obtained but it lacked aggressiveness and would be unlikely to become established in nature.

10.4 Transferring Rust Resistance to Wheat from Related Species

Because the relatives of wheat are tremendously diverse, transferring genes to wheat can range from simple to very difficult. As a consequence, the methods used vary tremendously as well. Over the years, as more experience has been gained and new methods developed, ever wider crosses have become possible.

The probability of success in transferring a character wheat from a related species depends basically on three factors:

1. The crossability of the species and the fertility of the hybrid.
2. The ability of the chromosomes of the two species to pair.
3. The genetic complexity of the character.

10.4.1 Crossability of Species and Hybrid Fertility

Many species of *Triticum* and related genera are cross-compatible and a large number of hybrids have been produced (listed by Knobloch 1968). Crosses between durum or bread wheat and a closely related species often result in viable seeds. Although a few wider crosses also produce viable seeds, many others do not. In these cases it is essential to use embryo rescue to obtain hybrid plants (described in Sect. 10.4.2).

The genotypes of the parents are very important in determining the success of a cross. One combination of genotypes of two species may be compatible and another not. For example, Kerber and Dyck (1973) found that line RL5244 of *T. monococcum* was compatible with *T. turgidum* cultivar Stewart but not with Arnautka. Many attempts to cross wheat with barley failed until the right combination of genotypes was found.

Chinese Spring is well known for its crossability, particularly with rye, but also with other species. Falk and Kasha (1981) screened 56 diverse wheat lines for crossability with rye and obtained a range in seed set from 0–95.4%. There was a good correlation ($r = 0.75$) between crossability with rye and crossability with *Hordeum bulbosum*. Thomas et al. (1981) found that wheat lines that crossed readily with rye also set more seed when pollinated with pollen from other *Triticum* species and other genera including *Aegilops, Secale, Agropyron,* and *Elymus*, than did lines with low crossability. Thus, it appears that the same genes in wheat control crossability with many other related species and genera.

Lein (1943) identified two loci, *Kr1* and *Kr2*, that controlled crossability between wheat and rye. Riley and Chapman (1967) determined that *Kr1*, the more effective gene, is on chromosome 5B and *Kr2* is on 5A. The dominant alleles inhibit crossability and the genes have additive effects. These two loci are probably important in determining crossability of wheat with many species.

Reciprocal differences in interspecific and intergeneric crosses are not uncommon. It may be possible to make a cross in one direction but not in the other. Fedak (1980) and Shepherd and Islam (1981) reported that the cross Betzes barley ♀ × Chinese Spring wheat ♂ was easy to make, but the reciprocal was very difficult. However, Shepherd and Islam (1981) found that in later generations of the first cross pistilloidy became a major problem. Using the reciprocal cross, they were able to backcross F_1 plants to wheat and establish six of the seven possible addition lines carrying single barley chromosomes.

Reciprocal differences in crosses are almost certainly caused by cytoplasm-genome interactions. The Japanese workers have carried out extensive studies on these interactions, summarized in Tsunewaki (1980). If an alien species is used as the female in a cross to *T. aestivum* and the hybrids as females in backcrosses to *T. aestivum*, the effect is to transfer the *aestivum* chromosomes into the alien cytoplasm. Various problems, often including male sterility, can result. Any interspecific cross should always be attempted in both directions.

Fairly commonly interspecific or intergeneric hybrids are male sterile but they may be at least partially female fertile. Backcrossing using pollen from the wheat parent may be successful. The male sterility may persist, particularly if the alien species was used as the female parent in the original cross. In this case the sterility probably results from the transfer of the wheat genome into an alien cytoplasm. If the hybrid produces any fertile pollen that can be used in a backcross to wheat the problem should be overcome. Alternatively, backcrossing to a number of wheat cultivars may find one that carries genes that restore fertility. If not, then the reciprocal cross may be the only answer.

Environmental conditions can affect the fertility of hybrids and they should always be grown under the best environmental conditions available. Strong, vigorous plants are more likely to be fertile than poor ones. Often hybrids will do best under natural field conditions with ample moisture and light.

10.4.2 Using Colchicine to Produce Amphiploids

In many hybrids between wheat and its relatives, sterility is caused by the failure of chromosome pairing. In these cases doubling the chromosomes to produce an amphiploid may restore fertility. The basic procedure for doubling hybrids in the Triticinae involves treating young, washed seedlings at the one- to three-leaf stage in 0.05–0.30% colchicine for up to 72 h (Elliott 1958; Jensen 1974; Kaltsikes 1974). If only the crown is treated, then the higher dosages for longer periods can be used. The roots are more sensitive to colchicine and will survive only if the lower dosages are applied for shorter periods. Winkle and Kimber (1976) found that the addition of dimethyl sulfoxide (DMSO) increased the effectiveness of lower concentrations of colchicine on hybrids in the Triticinae. An effective treatment was immersing the roots and crown of washed seedlings in 0.05% colchicine plus 1.5% DMSO for 5 h. For barley haploids, Thiebaut and Kasha (1978) recommended treating the crowns of seedlings in 0.1% colchicine, 2% DMSO, 10 mg/1 GA_3 and 0.3 ml/1 Tween 20 for 5 h at 22°C, in light. Following any treatment, plants should be washed several times in water to remove the colchicine, and then given good growing conditions.

10.4.3 Embryo Rescue

In many interspecific and intergeneric hybrids, the embryos appear to develop normally for a number of days and then show signs of degeneration. This probably results from a breakdown of the endosperm and its inability to nourish the embryo. If the embryos have developed for at least 8–10 days, often they can be removed, cultured, and induced to produce plants.

Research on embryo culture began in the early 1900s and many studies were done on culture media and on the effects of environment (reviewed by Rappaport 1954). The most efficient medium differs for different species. For the Triticeae, modifications of Murashige and Skoog's (1962) medium are often used (Table 10.2) but several other media have also been used successfully (see Gamborg and Shyluk 1981). The medium, with about 0.7% agar added to solidify it, can be poured into test tubes which are then plugged with cotton, into screw top vials or into petri plates. The tubes, vials or plates are then autoclaved.

In doing embryo culture work, maintaining sterile conditions and sterile instruments is absolutely essential. If a sterile transfer chamber is available, it is a great help in reducing contamination. Hands and the working area can be wiped with 95% ethanol. Calcium hypochlorite (5–10%), mercuric chloride (0.1%), or ethanol (95%) can be used to sterilize spikes or seeds. Some workers surface sterilize the spikes before removing the seeds, while others sterilize only the seeds. The embryos are dissected out of the sterilized seeds using a fine needle or scalpel, usually under low

Table 10.2. Composition of the Murashige and Skoog (1962) medium (MS medium) commonly used for embryo culture in wheat. (Gamborg and Shyluk 1981)

Macronutrients	mg/l	mM
NH_4NO_3	1650	20.6
KNO_3	1900	18.8
$CaCl_2{\cdot}2H_2O$	440	3.0
$MgSO_4{\cdot}7H_2O$	370	1.5
KH_2PO_4	170	1.2
Micronutrients		
KI	0.83	5
H_3BO_3	6.2	100
$MnSO_4{\cdot}4H_2O$	22.3	100
$ZnSO_4{\cdot}7H_2$	8.6	30
$Na_2MoO_4{\cdot}2H_2O$	0.25	1.0
$CuSO_4{\cdot}5H_2O$	0.025	0.1
$CoCl_2{\cdot}6H_2O$	0.025	0.1
Fe-Versenate (EDTA)	43.0	100
Vitamins and hormones		
Inositol	100	
Nicotinic acid	0.5	
Pyridoxine·HCl	0.5	
Thiamine·HCl	0.1	
IAA	1–30	
Kinetin	0.04–10	
Sucrose	30,000	
pH	5.7	

magnification, e.g., a 10–25× dissecting microscope. They are then transferred to petri plates or agar slants using forceps. The general procedure for culturing embryos is as follows:

1. Cross the parents. Some workers then spray the spikes with 75 ppm GA_3 a few hours after pollination and again the next day. It is not clear whether this is effective or not.
2. Allow the embryos to grow as long as possible but remove them as soon as there is any sign of degeneration (browning, watery appearance). Embryos removed before 8–10 days are difficult or impossible to culture. At the appropriate time, remove the seeds from the spike. Sterilize them, for example using a 0.1% $HgCl_2$ solution for 2 min, and rinse in distilled water. Keep the seeds slightly moist.
3. Place the seed on sterile filter paper and cut out the embryo under a dissecting microscope using sterilized needles or scalpels. Place the embryo scutellar side down on the medium in tubes or petri dishes.
4. Germinate embryos in an incubator at 25–27°C in the dark for up to 14 days. Once germination has started, transfer the seedlings to a growth chamber at 25°C with 16 h light and 8 h dark periods. If the embryos do not germinate in 15–20 days, place them in a refrigerator at 2–5°C for 7 days and then return them to the incubator. The temperature shock may stimulate germination.

5. When the shoots are 2–5 cm in length and roots have developed, transfer the seedlings to sterile vermiculite or Hoagland's nutrient solution for further growth. Since the seedlings are very tender at this stage, it is a good idea to cover them with a glass beaker to maintain high humidity.

Different workers use various modifications of this general procedure.

10.4.4 Two Historical Transfers

In early breeding work it was observed that the tetraploids such as *T. turgidum* and *T. timopheevii* frequently showed good disease resistance. Several successful attempts were made to transfer their rust resistance to bread wheat.

Hayes et al. (1920) crossed Iumillo durum with Marquis bread wheat and, after several generations of selfing, selected out the stem and leaf rust-resistant line Marquillo. Although Marquillo never became a cultivar, it was a parent of the very important cultivar Thatcher. Thatcher has several genes for specific resistance to stem rust and is thought to also have resistance of the slow rusting type. However, it has not been a particularly successful parent in simple crosses. It has been used extensively as a recurrent parent in backcrossing programs by Dr. A.B. Campbell, Agriculture Canada Research Station, Winnipeg. Its derivatives such as Manitou, Neepawa, Katepwa, and Columbus are very important cultivars in Canada.

In a similar program, McFadden (1930) crossed Yaroslav emmer with Marquis and produced the leaf and stem rust-resistant lines Hope and H44-24. Neither was successful commercially (although Hope was grown on a few thousand acres), but both have been widely used as sources of rust resistance in crosses. Hope has several genes for specific seedling resistance (*Sr7b, Sr9d, Sr17, Sr18*) plus *Sr2* for adult plant resistance, and may have additional genes having smaller effects on adult plant resistance.

In both these early transfers of resistance from tetraploids, it is not surprising that satisfactory commercial types were not obtained since no backcrosses to bread wheat were made. The recovery of hexaploids depended on selection for bread wheat characters. As would be expected, the lines derived from the crosses still retained some deleterious characters from the tetraploids.

10.4.5 Transfers Involving Genes on Homologous Chromosomes

The difficulty of making gene transfers from wild relatives to durum or bread wheats depends in part on whether the genes to be transferred are on genomes homologous to the A, B, or D genomes or are on genomes that are only homoeologous. Transfers involving homologous chromosomes are comparatively easy while those involving homoeologous chromosomes are much more difficult and require special techniques.

10.4.5.1 Transfers from AB Tetraploids

The wild tetraploids carrying the AB genomes, now designated *T. turgidum* var. *dicoccoides*, but often called *T. dicoccoides* or *T. dicoccum,* cross readily with durum wheat. Chromosome pairing is relatively normal and the hybrids are fertile. Gene transfers by backcrossing are easy. Few papers have been published on the transfer of rust resistance to durum wheat from wild tetraploids. However, Gerechter-Amitai and Grama (1974) crossed a very stripe rust-resistant selection of *T. dicoccoides* with durum wheat and found that resistance was simple. Considerable work has been done transferring high protein from *T. turgidum* var. *dicoccoides* to durum wheat (Avivi et al. 1983).

The AB tetraploids generally cross readily with bread wheat, the seeds are viable and the hybrids are usually vigorous and partially fertile, although fertility may be low in crosses with the wild tetraploids. However, there is increasing evidence that the success of crosses can vary depending on the genotypes used and the direction of the cross. The bread wheat parent is commonly used as the female. However, when Grama and Gerechter-Amitai (1974) transferred stripe rust resistance from *T. dicoccoides* (*T. turgidum*) G-25 to bread wheat, they obtained seed set only when the bread wheat parent was used as the male. Thus, in any program to transfer resistance from AB tetraploids to bread wheat, several combinations of genotypes and reciprocal crosses should be tried.

Hybrids between the AB tetraploids and bread wheat have a maximum of 14 pairs of chromosomes and at least 7 univalents. To transfer rust resistance, the hybrids should be backcrossed several times to the bread wheat parent to recover 21 pairs of chromosomes and the bread wheat phenotype. Although gametes with a range of chromosome numbers are produced by F_1 plants, those that have 14 or 21 chromosomes or numbers close to them are more likely to function. The fewer the backcrosses that are made (perhaps one or two), the greater is the chance of recovering genes for other useful characters such as increased yield or protein content. However, a larger number of backcrosses will reduce the possibility of transferring deleterious genes and increase the probability of recovering the genotype of the recurrent parent.

Any transfer will involve a block of genes linked to the gene for rust resistance. If homology between the alien and wheat chromatin is incomplete, the linkage block may not be easily broken up by crossing over. Not uncommonly the linkage block contains genes for deleterious characters, such as tenacious glumes which prevent free threshing. When stem rust resistance was transferred from Gaza durum to Bobbin bread wheat, a pollen-killer gene was also transferred (Loegering and Sears 1963) which resulted in distorted ratios in certain crosses. However, the locus has no obvious effect when either allele is homozygous.

Another major problem can be the failure of genes to be expressed when transferred from a tetraploid to a hexaploid, apparently as a result of inhibitors in the D genome. Kerber (1983) was unsuccessful in attempts to transfer leaf rust resistance from Stewart 63 durum wheat to Canthatch and Marquis bread wheat. He then crossed Stewart 63 with *T. tauschii* (DD) and doubled the chromosomes to produce the amphiploid. It was susceptible to leaf rust, showing that resistance was suppressed by a gene in the D genome. Kerber and Green (1980) found that Tetra

Canthatch, an AB tetraploid derived from Canthatch (ABD), was resistant to some races of stem rust to which Canthatch was susceptible. They concluded that Canthatch carries a gene on a D genome chromosome that inhibits a resistance gene on an A or B genome chromosome. The gene was located on the long arm of chromosome 7D (7DL). The suppressor is apparently very common since Kerber (1983) found it in six bread wheat cultivars and six synthetic hexaploids (ABD) that he tested. It would be interesting to know whether the presence of suppressors has prevented the transfer of other genes for rust resistance from AB tetraploids to bread wheat. However, it is not known whether suppressors of resistance are common or not.

The literature does not indicate that there has been much interest in transferring genes from bread to durum wheat, although it should be feasible to do so if the genes are on A or B genome chromosomes. However, the gene for semidwarfism, *Rht1* on chromosome 4A, has been transferred from bread to durum wheat in several breeding programs. The gene *Sr7a* which gives resistance to race 15B of stem rust, was transferred from bread to durum wheat by E.R. Kerber (personal communication 1984).

10.4.5.2 Transfers from Einkorn Wheat (AA) to Durum and Bread Wheat

Although einkorn wheats (*T. monococcum*) often cross readily with bread wheat, viable seeds do not normally result and embryo culture must be used. Different crosses appear to differ in the ease with which their embryos can be cultured. An alternative is to first cross the einkorn wheat to a durum and, perhaps, backcross to durum. The gene or genes being transferred may be valuable in durum wheat, in any case. The tetraploid can then be crossed and backcrossed to a bread wheat to complete the transfer. This procedure was used by Gerechter-Amitai et al. (1971) and The (1973) to transfer *Sr21* and *Sr22* to bread wheat.

Several problems have arisen in the transfer of genes from einkorn wheats. Kerber and Dyck (1973) found that some crosses between einkorn and durum wheats are compatible while others are not. Einkorn line RL5244 was compatible with the durum cultivar Stewart but not with Arnautka. The presence of suppressors of rust resistance in the D genome has already been noted. In some cases, in transferring stem rust resistance from einkorn to durum to bread wheat, the resistance has not been suppressed but it has progressively weakened. In one transfer, a gene which conditioned an IT 0;1 in an einkorn accession, conditioned an IT 1^{+} when transferred to a durum and an IT 2 when transferred to a bread wheat (Kerber and Dyck 1973). Nevertheless, einkorn wheats are a relatively easily exploitable source of genes for use in durum and bread wheats.

10.4.5.3 Transfers from *T. tauschii* (DD) to Bread Wheat

Synthetic hexaploids (AABBDD) have frequently been produced by crossing *T. turgidum* (AABB) with *T. tauschii* (DD) and doubling the chromosome numbers. Plump seeds can be obtained when the tetraploid is used as the female parent, although the reciprocal cross fails. The hybrids are largely sterile but colchicine can be used to produce amphiploids. The synthetic hexaploid can then be crossed with bread wheat for the transfer of genes from *T. tauschii*. Chromosome pairing in the hybrids is usually good.

Kerber and Dyck (1969) and Dyck and Kerber (1970) used this procedure to transfer two genes for seedling resistance to leaf rust and one for adult plant resistance to bread wheat from *T. tauschii*. The gene for adult plant resistance proved to be linked to a gene for tenacious glumes (non-free threshability).

Somewhat surprisingly, Merkle and Starks (1985) found that *T. tauschii* could be used as a parent in a direct cross with bread wheat. Embryo culture was not necessary and the hybrids could be backcrossed to bread wheat. Thus, this provides an alternative method of gene transfer.

Although *T. tauschii* does not appear to be as good a source of genes for rust resistance as some other species, it does have some useful resistance. And it does have special interest as the donor of some of the special breadmaking properties of bread wheat.

10.4.5.4 Transfers from *T. timopheevii* (AG) to Bread Wheat

T. timopheevii is particularly noted for its resistance to diseases, including the rusts. It carries the A genome and a second genome that has sometimes been considered to be a modified B genome, but is now generally thought to be a distinct genome designated G. Genes in the A genome should be easier to transfer to bread wheat than genes on the G genome chromosomes, which are less likely to pair with wheat chromosomes.

T. timopheevii can be crossed fairly easily with bread wheat, although the degree of difficulty varies with different crosses. The best seed is obtained when *timopheevii* is used as the male parent. The hybrids are much less fertile than those between *T. turgidum* and bread wheat, but seeds can usually be obtained from a backcross using pollen from bread wheat. However, male sterility and even complete sterility is often a continuing problem in backcrosses. This is not surprising considering that *T. timopheevii* has been an important source of cytoplasmic male sterility for hybrid wheat breeding, and also of fertility-restoring genes.

Despite the fertility problems, several transfers of disease resistance from *T. timopheevii* to bread wheat have been made. Allard and Shands (1954) crossed a *T. timopheevii* accession with Illinois No. 1/Chinese Spring and backcrossed twice. They were successful in transferring resistance to stem rust, leaf rust, and powdery mildew. All three resistances appeared to be linked and stem rust resistance seemed to be controlled by two linked genes. Later, in the F_2 populations from crosses between the *timopheevii* derivative C.I. 12633 and five susceptible cultivars, Nyquist (1962) obtained from 7.1 to 26.6% of stem rust susceptible plants. He concluded that only one gene was segregating but that differential fertilization occurred in different genetic backgrounds. The same gene was also transferred by Pridham (1939). It was originally designated *SrTt1*, but now has been given the symbol *Sr36*. It has been used in a number of cultivars including Idaed 59 and Arthur 71 in the U.S.A. and Mengavi, Timgalen, and Timvera in Australia.

One other gene for stem rust resistance, *Sr37* (= *SrTt2*), was transferred from *T. timopheevii* using a race virulent on C.I. 12632 (*Sr36*) (McIntosh and Gyarfas 1971).

McIntosh and Gyarfas (1971) compared the reactions to seven races of stem rust of 22 accessions of *T. timopheevii* and three lines of bread wheat carrying genes for stem rust assistance that had been transferred from *timopheevii*. They concluded that *T. timopheevii* carries additional genes for stem rust resistance. There is no doubt that various accessions of *T. timopheevii* carry useful genes for rust resistance that have not yet been exploited. However, because of the difficulty of making transfers, it may be better to attempt to use resistances from other species.

T. timopheevii has also been crossed to durum wheat, particularly in Russia. Pandey and Rao (1983) found that when F_1 plants were backcrossed to durum wheat at low temperatures, seed set was poor. Later in the spring when temperatures were higher, seed set was good. Deodikar et al. (1979) successfully transferred leaf rust resistance from *T. timopheevii* to durum wheat.

10.4.6 Transfers Involving Species with Genomes Homoeologous to the A, B, and D Genomes

Many of the more distant relatives of wheat are known to carry resistance to the rusts. Since their chromosomes are not homologous with the A, B, or D genomes, their genes for resistance cannot be transferred by normal crossing over. Special methods involving either irradiation or the induction of homoeologous pairing are required.

A surprising range of species and genera can be crossed with wheat, with varying degrees of difficulty. Embryo culture is usually necessary to obtain the polyhaploid hybrid plants. Chromosome doubling to produce amphiploids is often used to improve fertility.

10.4.6.1 Production of Addition and Substitution Lines

The procedures for transferring genes to wheat from its more distant relatives work best with major genes whose effects are easy to measure. Fortunately, this is the case for many rust resistances that are of interest, although other resistances may be polygenic and difficult to transfer.

Often, but not always (as will be discussed later), the first step is to add the chromosome carrying the resistance gene to the normal wheat complement, or to substitute it for a wheat chromosome. An addition line can be produced by backcrossing a polyhaploid F_1 or an amphiploid to the wheat parent and selecting for resistance. Usually resistance proves to be due to a single gene. After several backcrosses, plants that are 21″ + 1′A (alien chromosome) are selfed to produce a disomic addition, 21″ + 1″A. Disomic addition lines are often fairly normal in appearance and vigor, and reasonably stable. However, the occasional failure of the alien chromosomes to pair at meiosis and be included in a nucleus results in their gradual loss. Normal 21-chromosome gametes probably have a competitive advantage over gametes with 21″ + 1′A in most cases. However, several cases have been reported of preferential transmission of alien chromosomes through the gametes (e.g., Endo 1982; Miller et al. 1982; Kibirige-Sebunya and Knott 1983).

Sometimes monosomic substitutions, $20'' + 1' + 1'A$, occur naturally during backcrossing and a disomic substitution, $20'' + 1''A$, can be produced by selfing. Usually the substitution involves a homoeologous wheat chromosome. Substitution lines can also be produced by crossing an addition line with a female monosomic for the chromosome which is to be replaced. The progeny will be either $21'' + 1'A$ or $20'' + 1' + 1'A$. Selfing the latter type may result in a disomic substitution, particularly if the alien chromosome is homoeologous to the wheat chromosome being replaced. It is a major advantage to know the homoeology of the alien chromosome and to make crosses only with the three wheat lines monosomic for the homoeologous wheat chromosomes. As an alternative to selfing, the double monosomic plants ($20'' + 1' + 1'A$) may be used as males in a backcross to the monosomic parent. They should produce about equal numbers of $20' + 1'$ and $20' + 1'A$ pollen. If the alien chromosome substitutes well for the missing wheat chromosome, some $20' + 1'A$ pollen should function and give rise to $20'' + 1'A$ plants. Selfing of these plants will produce the disomic substitution.

Sears (personal communication 1986) has suggested another procedure for producing alien substitutions. Plants that are $20'' + 1' + 1'A$ are pollinated with pollen from the alien addition line ($21'' + 1''A$). About 3/16 of the eggs should be $20' + 1'A$ and another 3/16 $20' + 1'$. If there is no selection against the alien chromosome in the eggs, then about 3/8 of the progeny should show $21'' + 1'$ (half $20'' + 1'' + 1'A$ and half $20'' + 1' + 1''A$, but indistinguishable). On selfing, the latter type will not segregate for resistance and should produce the disomic substitution ($20'' + 1''A$) even if male transmission of the alien chromosome is poor.

Often substitutions can be produced only for homoeologous chromosomes. Substitutions for nonhomoeologous chromosomes are usually inviable or very weak. The vigor of substitution lines is determined by the degree of compensation of the alien chromosome for the wheat chromosome. Homoeologous substitutions may be essentially normal in vigor. They are also stable, since pairing is usually good. If an alien pair of chromosomes fails to pair, the loss of a chromosome will result in a gamete that has 20 chromosomes and may fail to function, particularly in pollen. Thus, substitution lines have some potential as commercial cultivars as long as the substituted chromosome does not carry deleterious genes. Chromosome 1R of rye, which carries resistance to stem, leaf, and yellow rust, and to powdery mildew, has replaced chromosome 1B in a number of successful European wheat cultivars (Mettin et al. 1973; Zeller 1973).

10.4.6.2 Spontaneous Translocations

When an alien species is crossed and backcrossed to wheat, spontaneous translocations between alien and wheat chromosomes occur occasionally. During the backcrossing, univalent alien and wheat chromosomes will frequently be present. Univalent chromosomes often misdivide at the centromere and arms of two different chromosomes may then rejoin, occasionally giving rise to a chromosome with one arm from a wheat chromosome and one from an alien chromosome. Translocations involving only part of a chromosome arm may also occur.

In addition to the substitution of rye chromosome 1R for wheat chromosome 1B in several European cultivars (mentioned above), several cultivars carry 1B/1R translocations (Mettin et al. 1973; Zeller 1973). The Veery lines developed in the CIMMYT program in Mexico also carry a 1B/1R translocation derived from the winter wheat, Kavkaz (CIMMYT 1985). They have proven to be very high-yielding in international tests, as well as having good resistance to several diseases.

In a program in Oklahoma, Smith et al. (1968) transferred leaf rust resistance from a *Triticum-Agropyron elongatum* derivative to a wheat cultivar, Agent. Resistance was controlled by a single dominant gene (*Lr24*) and involved a natural translocation to wheat chromosome, 3D. Agent also carried a gene (*Sr24*) for resistance to stem rust. Both genes were transferred to a number of winter wheats in the U.S.A. Virulence on *Lr24* appeared in leaf rust in the field but *Sr24* has remained effective against stem rust.

It may be possible to use spontaneous translocation to transfer genes to wheat chromosomes from alien chromosomes. Sears (1972a, 1981) suggested the production of plants that carry monosomic substitutions (i.e., $20'' + 1' + 1'A$). If disomic substitutions are available, they can be crossed to normal wheat plants to produce the monosomic substitution. Misdivision of the wheat and alien univalents can result in spontaneous translocations, although the frequency will be low. Zeller (1981) produced monosomic substitutions of rye chromosomes by crossing Chinese Spring monosomics with Chinese Spring wheat-rye addition lines. Two wheat-rye chromosome translocations resulted. Jan et al. (1981) crossed an addition line carrying a pair of chromosomes from *Elytrigia pontica* with Sonora 64. After 12 generations of selfing and selecting for blue aleurone color carried on the *Elytrigia* chromosome, one of 20 blue lines proved to carry a translocation. Several lines of evidence indicated that the translocation involved homoeologous arms of wheat chromosome 4A and *Elytrigia* chromosome 4el.

The discovery that frequent chromosome aberrations, including breakages and interchanges, occur in tissue cultures suggests another possible mechanism for producing spontaneous translocations (Lapitan et al. 1984). Tissue cultures of alien addition or substitution lines or even of interspecific or intergeneric hybrids can be grown, and plants regenerated from them and studied cytologically. Regeneration of plants in wheat is difficult but can be done.

10.4.6.3 The Use of Radiation to Produce Translocations

The various genomes of wheat and its relatives are thought to have arisen from a common genome in the distant evolutionary past. Nevertheless, many of the genomes in the relatives have differentiated sufficiently that their chromosomes rarely pair with the A, B, and D genome chromosomes of bread wheat. And, of course, the A, B, and D genome chromosomes do not pair with one another.

The failure of homoeologous chromosomes to pair makes the transfer of rust resistance and other characters to wheat more difficult. Sears (1956) pioneered a method using ionizing radiation to produce translocations between homoeologous chromosomes. He wanted to transfer leaf rust resistance from *Aegilops umbellulata* (*T. umbellulatum*, UU) to bread wheat. Only inviable seeds were obtained in the cross between *T. umbellulatum* and *T. aestivum*. Therefore, Sears (1956) crossed *T. umbellulatum* with *T. dicoccoides* (*T. turgidum* var. *dicoccoides*, AABB), produced the amphiploid, and crossed it with Chinese Spring wheat. The resulting F_1 plants were both male and female fertile and were used as pollen parents in a backcross to Chinese Spring. After one further backcross, resistant plants with 21″ plus an added *umbellulatum* chromosome were obtained, including one carrying an isochromosome for the long arm of the *umbellulatum* chromosome, the arm carrying resistance. Plants carrying the isochromosome were irradiated with X-rays, mostly at 1250 or 1500r, when the first spikes were entering meiosis. Pollen from these plants were used to pollinate Chinese Spring. Since the *umbellulatum* chromosome was deleterious and showed low transmission through the pollen, it was expected that most resistant progeny would carry a translocation. Of 6091 plants, 132 were rust resistant and 40 proved to carry a translocation when examined cytologically. One line was designed Transfer, and its gene for resistance, *Lr9*, was used in several cultivars in the United States but was quickly overcome by the pathogen.

Following Sears' (1956) development of the procedure, a number of other workers used similar methods to transfer rust resistance, particularly from *Agropyron* species (= *Thinopyrum* and *Elytrigia*) to wheat (reviewed by Knott 1971b, 1987). Various types of radiation, including gamma rays and neutrons, were used on either seeds or plants. The procedures are slow and require a good deal of cytological work. Chromosome breakage and reunion presumably occur at random, so most of the translocations are likely to involve nonhomoeologous chromosomes and have deleterious effects. However, in two separate programs, Knott (1961) and Sharma and Knott (1966) transferred stem rust and leaf rust resistance, respectively, from an *Agropyron elongatum* chromosome to a wheat chromosome. Seven and five translocations were produced in the two programs. In heterozygotes, transmission of the translocations was usually normal through the eggs but reduced through the male gametes. In each case, one translocation was transmitted normally through both male and female gametes. Both proved to involve homoeologous wheat and *Agropyron* chromosomes. This suggests that translocations involving homoeologous chromosomes are more likely to survive and be transmitted through the gametes.

The discovery of the genetic control of homoeologous pairing in wheat provided a basis for the development of better methods of producing alien transfers. In the future, radiation will most likely be used to produce translocations between wheat and alien chromosomes only if the chromosomes are so dissimilar that homoeologous pairing cannot be induced.

10.4.6.4 The Use of Induced Homoeologous Pairing

The genetic control of homoeologous pairing in wheat was discovered almost simultaneously by Okamoto (1957), Sears and Okamoto (1958), Riley (1958) and Riley and Chapman (1958). Okamoto crossed Chinese Spring monotelosomic 5B with an AADD amphiploid. Compared to 35-chromosome plants, the 34-chromosome progeny had many fewer univalents (8.24 compared to 23.82) and much more pairing including multivalents with up to seven chromosomes. Riley (1958) and Riley and Chapman (1958) obtained three types of nulli-haploids (20 chromosomes) in the cultivar Holdfast. One nulli-haploid had 4.2 bivalents and 0.8 trivalents per cell, many more than the other two (0.7–0.9 bivalents per cell). The missing chromosome was later identified as 5B. Thus, both lines of evidence indicated that chromosome 5B carries a gene (now designated *Ph1*) which essentially prevents homoeologous pairing.

Although the *Ph1* locus is the major contributor to the control of homoeologous pairing, a number of other genes, primarily on chromosomes of homoeologous groups 2, 3, and 5, are involved.

Three basic procedures have been developed to manipulate the *Ph1* gene and induce pairing between alien chromosomes and their wheat homoeologs:

1. The use of crosses involving aneuploid types to produce plants carrying an alien chromosome or chromosomes, but lacking chromosome 5B.
2. The use of crosses with species carrying a suppressor of *Ph1*.
3. The use of a mutant of the *Ph1* or some other locus affecting homoeologous pairing.

10.4.6.4.1 The Development of Plants Lacking Chromosome 5B

If an alien species can be crossed directly with bread wheat, plants monosomic for chromosome 5B can be used as female parents in crossed with it. About 75% of the progeny will lack chromosome 5B, and pairing among homoeologous chromosomes should occur. The frequency of homoeologous pairing should be high, since none of the chromosomes has homologs with which to pair. However, the plants may show considerable sterility, although they often can be used as female parents in a backcross to wheat.

Direct crosses between a monosomic 5B line and an alien species seem to have been used rarely. However, Riley (1966) transferred genetic material from *Ae. bicornis* (*T. bicorne*) to wheat in this manner, although the transfer did not involve rust resistance. Joshi and Singh (1979) tried the procedure by crossing monosomic 5B of Pb.C591 wheat with Russian rye. Although no 27-chromosome plant was obtained, one of three 28-chromosome plants showed extensive pairing. The 28-chromosome plants were expected to carry chromosome 5B. However, the exceptional plant may have lacked 5B but been disomic for another chromosome, or it may have carried a *ph* mutant. It was backcrossed twice to Pb.C591 and plants with resistance to all three rusts were obtained.

An alternative method for producing plants lacking chromosome 5B involves the use of a nullisomic 5B tetrasomic 5D line of Chinese Spring. Different crossing schemes can be used, depending on the stocks available. For example, Sears (1973) crossed monosomic 5B with a substitution line in which an *Agropyron* chromosome carrying leaf rust resistance had been substituted for chromosome 7D (Fig. 10.1). The monosomic 5B progeny were then crossed to nullisomic 5B tetrasomic 5D. Resistant plants nullisomic for 5B were crossed and backcrossed to a normal susceptible and 12 translocations were obtained. Four translocations showed more than 50% transmission through male gametes of heterozygotes, while three showed

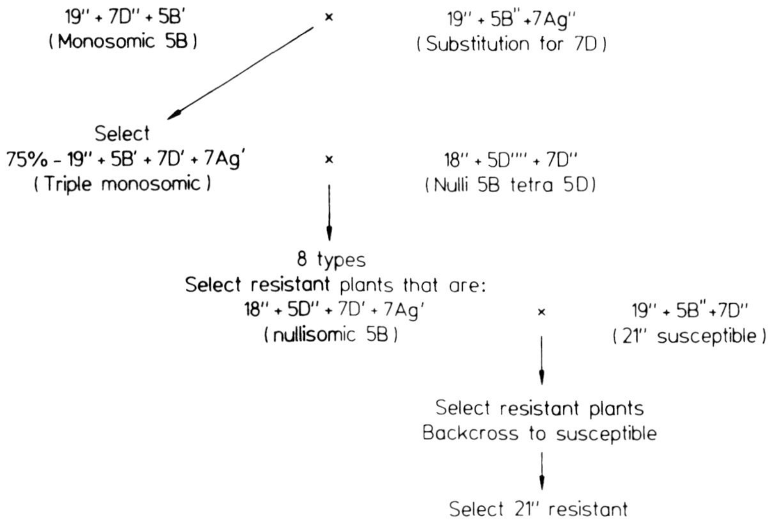

Fig. 10.1. The use of a nullisomic 5B tetrasomic 5D line to induce homoeologous pairing between an *Agropyron* chromosome and a wheat chromosome

significantly less. All appeared to involve wheat chromosome 7D. The differences in degree of compensation in the male gametes presumably resulted from differences in the chromosome segments that were exchanged.

In a similar program involving another *Agropyron* chromosome (3Ag) carrying leaf rust resistance, Sears (1973) produced 20 translocations. Fifteen involved wheat chromosome 3D but five involved other chromosomes, presumably 3A or 3B.

The low fertility of the triple monosomic plants can be a problem in this procedure but this has not prevented its use in several programs.

10.4.6.4.2 Crosses with Species Carrying a Suppressor of Ph1

Specific genotypes of *Ae. speltoides* (*T. speltoides*) and *Ae. mutica* (*T. tripsacoides*) have been found to inhibit the effect of *Ph1* to varying degrees. Since both species can be crossed with wheat using embryo rescue, and the hybrids backcrossed to wheat, direct transfers of genetic material from them to wheat can be made. Fertility is usually low in the hybrids but some seed can be obtained when they are used as female parents in backcrosses. Homoeologous pairing should be frequent since no homologous chromosomes are present (21'ABD + 7'S). Dvořák (1977) crossed five accessions of *T. speltoides* with wheat and successfully transferred leaf rust resistance from each. After extensive backcrossing, resistant lines derived from two of the accessions look promising for commercial use. Lines from the remaining three accessions have been generally low-yielding and it is clear that deleterious genes are tightly linked to genes for resistance. Even transfers of homoeologous segments from fairly closely related species can have deleterious effects and it is evident that linkages within the segments are different to break.

Procedures have also been developed in which *T. speltoides* and *T. tripsacoides* are used to transfer genes from other species to wheat. One procedure is to cross an alien addition line carrying the desired gene with a genotype of *T. speltoides* that results in a high frequency of pairing. The hybrid is then backcrossed to wheat and the desired character selected for. If the character is rust resistance, then, if possible, a genotype of *T. speltoides* (or *T. tripsacoides*) that is susceptible should be used. Otherwise the source of the rust resistance that is transferred will be uncertain.

Riley et al. (1968a) successfully transferred yellow rust resistance from *Ae. comosa* (*T. comosum*) to wheat in this way and the resulting line was called Compair. The recombinant chromosome carried a piece of the long arm of wheat chromosome 2D transferred to *comosum* chromosome 2M.

10.4.6.4.3 Crosses Involving Ph Mutants

Once it was determined that pairing among homoeologous chromosomes in wheat is prevented by the gene *Ph1*, the next logical step was to produce *ph* mutants. Both Okamoto (1966) and Sears (1966 unpublished, see Sears 1977) attempted to produce mutants by irradiating emasculated wheat spikes and then pollinating them with rye pollen. If a mutation occurred, the resulting plants should have shown multivalents at meiosis. Over 3% of the progeny showed high pairing but the plants were highly sterile and no mutants were recovered even after attempts to double the chromosome number. In any case, the mutants were probably long deletions (Sears 1984).

Wall et al. (1971a,b) used ethyl methanesulfonate to produce a pairing mutant that appeared to be on chromosome 5B. The mutant resulted in substantially less pairing than did the absence of chromosome 5B. Sears (1984) showed that the gene was actually a mutant at a locus on chromosome 3D and designated it *ph2b*. He also determined that there was 2.5% crossing over between *ph2b* and the centromere.

Sears (1977) obtained a *Ph1* mutant using a procedure involving a chromosome 5B carrying a marker gene. Spikes of normal Chinese Spring were irradiated with 500r of X-rays just prior to anthesis. Pollen from them was put on plants monosomic for a chromosome 5B having all or part of its short arm replaced by a segment of rye chromatin carrying *Hp*, a gene for hairy neck. Thus, any progeny showing hairy neck received the monosome and were normal euploids. They were discarded. The nonhairy necked plants carried only an irradiated 5B from Chinese Spring, unless it had been lost and the plants were nullisomic 5B. Six hundred and three plants were nonhairy and reasonably fertile. Of these, 438 were tested in various ways for homoeologous pairing, including crosses to *T. kotschyi* (*Ae. variabilis*) females. One high-pairing mutant was obtained and designated *ph1b*. An intermediate mutant proved to be on chromosome 3D and was designated *ph2a* (Sears 1982, 1984). All of the mutants are probably deletions.

The mutants, particulary *ph1b*, cause pairing among homoeologous wheat chromosomes which results in unbalanced gametes and some sterility. They are probably best maintained as heterozygotes. Sears has produced a line in which *ph1b* is on a telosome. In the progeny of a heterozygote, homozygous plants can be selected by root tip counts, since they will be ditelosomic.

The *ph* mutants have not been used extensively as yet but there is some evidence of a problem that may arise. Sharma and Gill (1986) crossed five tetraploid and one hexaploid species with both Chinese Spring and a *ph1* mutant line. They were able to backcross the F_1 plants from the Chinese Spring crosses but obtained no seed on the F_1 plants from the crosses involving the mutant. Successful backcrossing often depends on the production of restitution gametes and the *ph* gene may affect this. Sharma and Gill (1986) suggest making the first cross with a highly crossable line such as Chinese Spring and then making a backcross to a *ph* mutant line. It may in fact be easier to produce an addition or substitution line carrying the desired resistance and use it as the starting point. Kibirige-Sebunya and Knott (1983) did this in transferring stem rust resistance to wheat from a substitution line carrying a pair of *Agropyron* chromosomes. The procedure is very similar to that using nullisomic 5B tetrasomic 5D (Fig. 10.1), except that a line carrying a *ph* mutant is substituted for the nullisomic 5B tetrasomic 5D. In fact, Kibirige-Sebunya and Knott (1983) used both types of cross. Both the entire *Agropyron* chromosome and the derived translocations showed very high preferential transmission through the female gametes.

10.5 The Results of Transfers of Rust Resistance from Alien Species to Wheat

The early work in this area involved the transfer of rust resistance to wheat from closely related tetraploids. As already mentioned, two of the best-known involved the transfer of rust resistance from Iumillo durum to Marquillo and Thatcher, and from Yaroslav emmer to Hope and H44-24. Thatcher became widely grown in North America and was also grown in other spring wheat areas such as China and Russia. However, it was never successful as a parent except where it was used as a recurrent parent in backcrosses. On the other hand, Hope and H44-24 were never grown to any extent but were widely used in breeding for resistance to stem rust.

Transfers of *Sr36* (*SrTt1*) from *T. timopheevii* were made in Australia (Pridham 1939) and the U.S.A. (Allard and Shands 1954) and used in cultivars in both countries. However, *Sr36* did not provide resistance to race 15B, which was important in North America, and it was quickly overcome in Australia when it was released in the cultivar Mengavi. Curiously, Luig (1979) found that Australian race 126 showed a high natural mutation rate to virulence on *Sr36,* but attempts to produce mutations in other races using EMS failed.

The transfers from various other close relatives of wheat appear not to have been used in commercial cultivars.

Extensive work has been done on the transfer of rust resistance to wheat from its more distant relatives such as rye and the wheatgrasses (formerly *Agropyron* spp., now *Elytrigia, Thinopyrum,* etc.) (Table 10.3). A number of transferred genes have been used extensively in commercial cultivars, but in all but one case (*Sr26*) they have been overcome by virulent mutants in the pathogen.

A number of other transferred genes have not yet been used in commercial cultivars. Most translocations probably involve the substitution of a segment of alien chromatin for a terminal segment of wheat chromatin. Deleterious effects may arise either because of the loss of desirable wheat genes or, more probably, from the effects of deleterious genes linked to the gene for rust resistance. Since the alien

Table 10.3. Genes for rust resistance that have been transferred to wheat from distant relatives

Alien species	Pathogen and gene	Method used	Cultivars produced	Reference
Aegilops umbellulata	*P. recondita Lr9*	X-rays	Riley 67, Arthur 71, etc.	Sears (1956)
Agropyron elongatum	*P. graminis Sr26*	Gamma rays	Eagle, Kite, etc.	Knott (1961)
Secale cereale	*P. graminis Sr27*	X-rays	–	Acosta (1963)
Secale cereale	*P. recondita Lr25*	X-rays, neutrons	–	Driscoll and Jensen (1964)
Agropyron elongatum	*P. graminis Sr25, P. recondita Lr19*	X-rays, neutrons	–	Sharma and Knott (1966)
Agropyron elongatum	*P. graminis Sr24, P. recondita Lr24*	Spontaneous translocation	Agent, etc.	Smith et al. (1968)
Secale cereale	*P. striiformis Yr9 P. recondita Lr26 P. graminis Sr31*	Spontaneous translocation	Aurora, Kavkaz, etc.	Mettin et al. (1973), Zeller (1973)
Aegilops comosa	*P. striiformis Yr8 P. graminis Sr34*	Cross with *Ae. speltoides*	–	Riley et al. (1968b)
Aegilops speltoides	*P. recondita* – 5 transfers	Direct cross with *T. aestivum*	–	Dvořák (1977)
Agropyron intermedium	All 3 rusts	Radiation	–	Wienhues (1979a, b)
Aegilops speltoides	*P. recondita Lr28*	Cross	–	McIntosh et al. (1982)
Agropyron elongatum	*P. graminis*	Nulli-5B tetra-5D, *ph* mutant	–	Kibirige-Sebunya and Knott (1983)

The species names used by the authors are given.

chromatin seldom pairs with wheat chromatin in the absence of *ph*, separation of the deleterious gene(s) from the gene for rust resistance by crossing over is unlikely. The induction of homoeologous pairing to produce transfers results in the exchange of homoeologous chromosome segments and should result in fewer deleterious effects than other procedures. When specific deleterious genes have been transferred, it may be possible to eliminate them by mutation. The transfers of *Lr19* and *Sr25* produced by Sharma and Knott (1966) carried a linked gene for yellow pigment. Knott (1980) used EMS to produce two mutants that lacked the gene for yellow pigment. They are probably deletions, and some preliminary data suggest that they may have some deleterious effects on yield.

In addition to the economic benefits that have resulted from some of the transfers, the data that have been obtained during their production have resulted in a wealth of information about the origin and evolution of wheat.

10.6 Future Prospects

The relatives of wheat provide a rich reservoir of genes of potential value in wheat breeding, including genes for rust resistance. These genes are most easily transferred to durum wheat from species that carry the A or B genomes and to bread wheat from species that carry the A, B, or D genomes. Transfers from these close relatives are also less likely to carry deleterious genes than those from more distant relatives. Even the closest relatives have been exploited to only a limited extent, mostly for simply inherited characters such as disease resistance, and almost not at all for more complex characters such as yeild or protein content. More attention should probably be given to rust resistance that is more complex in inheritance and, perhaps, more durable than simply inherited specific resistance. Such transfers will, however, be more difficult.

References

Acosta AC (1963) The transfer of stem rust resistance from rye to wheat. Diss Abstr 23:34–35

Alexander HM, Roelfs AP, Groth JV (1984) Pathogenicity associations in *Puccinia graminis* f. sp. *tritici* in the United States. Phytopathology 74:1161–1166

Allan RE (1987) Wheat. In: Fehr WR (ed) Principles of cultivar development, vol 2. Macmillan, New York, pp 699–748

Allan RE, Line RF, Peterson CJ, Jr., Rubenthaler GL, Morrison KJ, Rohde CR (1983) Crew, a multiline wheat cultivar. Crop Sci 23:1015–1016

Allard RW, Shands RG (1954) Inheritance of resistance to stem rust and powdery mildew in cytologically stable spring wheats derived from *Triticum timopheevi*. Phytopathology 44:266–274

Anderson MK, Williams ND, Maan SS (1971) Monosomic analysis of genes for stem rust resistance derived from Marquis and Reliance wheats. Crop Sci 11:556–558

Anderson RG (1961) The inheritance of leaf rust resistance in seven varieties of common wheat. Can J Plant Sci 41:342–359

Antonelli E, Daly JM (1966) Decarboxylation of indoleacetic acid by near-isogenic lines of wheat resistant or susceptible to *Puccinia graminis* f. sp. *tritici*. Phytopathology 56:610–618

Ashagari D, Rowell JB (1980) Postpenetration phenomena in wheat cultivars with low receptivity to infection by *Puccinia graminis* f. sp. *tritici*. Phytopathology 70:624–627

Ausemus ER, Harrington JB, Reitz LP, Worzella WW (1946) A summary of genetic studies in hexaploid and tetraploid wheats. J Am Soc Agron 38:1082–1099

Avey DP, Ohm HW, Patterson FL, Nyquest WE (1982) Three cycles of simple recurrent selection for early heading in winter wheat. Crop Sci 22:908–912

Avivi L, Levy AA, Feldman M (1983) Studies of high protein durum wheat derived from crosses with the wild tetraploid wheat, *Triticum turgidum* var. *dicoccoides*. In: Sakamoto S (ed) Proc 6th Int Wheat genetics Symp, Kyoto, pp 199–204

Baenziger PS, Wesenberg DM, Schaeffer DM, Galun E, Feldman M (1983) Variation among anther culture derived doubled haploids of Kitt wheat. In: Sakamoto S (ed) Proc 6th Int Wheat genetics Symp, Kyoto, pp 575–581

Baker EP (1966) Isolation of complementary genes conditioning crown rust resistance in the oat variety Bond. Euphytica 15:313–318

Barr R, Caldwell RM, Amacher RH (1964) An examination of vegetative recombination of urediospore color and virulence in mixtures of certain races of *Puccinia recondita*. Phytopathology 54:104–109

Bartoš P, Fleischmann G, Samborski DJ, Shipton WA (1969) Studies on asexual variation in the virulence of oat crown rust, *Puccinia coronata* f. sp. *avenae*, and wheat leaf rust, *Puccinia recondita*. Can J Bot 47:1383–1387

Basile R (1957) A diagnostic key for the identification of physiologic races of *Puccinia rubigo-vera tritici* grouped according to a unified numeration scheme. Plant Dis Rep 41:508–511

Beeson KE (1923) Common barberry and black stem rust in Indiana. Purdue Univ Dep Agric Ext Bull 118: 8 pp

Berg LA, Gough FJ, Williams ND (1963) Inheritance of stem rust resistance in two wheat varieties, Marquis and Kota. Phytopathology 53:904–908

Biffen RH (1905) Mendel's laws of inheritance and wheat breeding. J Agric Sci 1:4–48

Biffen RH (1911–12) Studies in the inheritance of disease resistance. J Agric Sci 4:421–429

Biffen RH (1931) The cereal rusts and their control. Trans Br Mycol Soc 16:19–37

Bingham J, Lupton FGH (1987) Production of new varieties: an integrated research approach to plant breeding. In: Lupton FGH (ed) Wheat breeding: its scientific basis. Chapman & Hall, London New York, pp 487–538

Borlaug NE (1959) The use of multilineal or composite varieties to control airborne epidemic diseases of self-pollinated crop plants. In: Jenkins BC (ed) Proc 1st Int Wheat genetics Symp, Winnipeg, pp 12–27

Bromfield KR (1961) The effect of postinoculation temperature on seedling reaction of selected wheat varieties to stem rust. Phytopathology 5:590–593

Browder LE (1971) Pathogenic specialization in cereal rust fungi, especially *Puccinia recondita* f. sp. *tritici*: concepts, methods of study, and application. USDA Tech Bull 1432: 51 pp

Browder LE (1972) Designation of two genes for resistance to *Puccinia recondita* in *Triticum aestivum*. Crop Sci 12:705–706

Browder LE (1973) Probable genotype of some *Triticum aestivum* 'Agent' derivatives for reaction to *Puccinia recondita* f. sp. *tritici*. Crop Sci 13:203–206

Browder LE (1980) A compendium of information about named genes for low reaction to *Puccinia recondita* in wheat. Crop Sci 20:775–779

Browder LE, Eversmeyer MG (1977) Pathogenicity association in *Puccinia recondita tritici*. Phytopathology 67:766–771

Browder LE, Eversmeyer MG (1980) Sorting of *Puccinia recondita: Triticum* infection-type data sets toward the gene-for-gene model. Phytopathology 70:666–670

Browder LE, Eversmeyer MG (1982) Sorting of infection-type data sets toward the gene-for-gene model: A reply. Phytopathology 72:458–460

Browder LE, Eversmeyer MG (1986) Interactions of temperature and time with some *Puccinia recondita: Triticum* corresponding gene pairs. Phytopathology 76:1286–1288

Browder LE, Young HC, Jr. (1975) Further development of an infection-type coding system for the cereal rusts. Plant Dis Rep 59:964–965

Browning JA, Frey KJ (1981) The multiline concept in theory and practice. In: Jenkyn JF, Plumb RT (eds) Strategies for the control of cereal disease. Blackwell, Oxford, pp 37–46

Browning JA, Frey KJ, McDaniel ME, Simons MD, Wahl IH (1979) The biologic of using multilines to buffer pathogen populations and prevent disease loss. Indian J Genet Plant Breed 39:3–9

Burdon JJ, Marshall DR, Ling NH, Gow DJS (1982) Isozyme studies on the origin and evolution of *Puccinia graminis* f. sp. *tritici* in Australia. Aust J Biol Sci 35:231–238

Busch RH, Kofoid K (1982) Recurrent selection for kernel weight in spring wheat. Crop Sci 22:568–572

Caldecott RS, Stevens H, Roberts BJ (1959) Stem rust resistant variants in irradiated populations – mutations or field hybrids? Agron J 51:401–403

Caten CE (1987) The concept of race in plant pathology. In: Wolfe MS, Caten CE (eds) Populations of plant pathogens: their dynamics and genetics. Blackwell, London, pp 21–37

Chapman V, Riley R (1966) The allocation of the chromosomes of *Triticum aestivum* to the A and B genomes and evidence on genome structure. Can J Genet Cytol 8:57–63

Chester KS (1946) The nature and prevention of the cereal rusts as exemplified in the leaf rust of wheat. Chron Bot, Waltham, MA

Cheung DSM, Barber HN (1972) Activation of resistance of wheat to stem rust. Trans Br Mycol Soc 58:333–336

Cimmyt (1985) CIMMYT research highlights 1984. International Maize and Wheat Improvement Centre, Mexico, DF

Cox DJ, Wilcoxson RD (1982) The relationship of the *Sr6* gene to slow rusting in wheat. Phytopathology 72:178–181

Craigie JH (1927) Discovery of the function of the pycnia of the rust fungi. Nature (London) 120:765–767

Craigie JH (1931) An experimental investigation of sex in the rust fungi. Phytopathology 21:1001–1040

Day PR (1974) Genetics of host-parasite interactions. Freeman, San Francisco

de Buyser J, Henry Y, Laur R, Lonnet P (1981) Utilisation de l'androgenèse in vitro dans des programmes de sélection du blé tendre (*Triticum aestivum* L.) Z Pflanzenzücht 87:290–299

Deodikar GB, Patil VP, Rao VS (1979) Breeding of Indian *durum* and *dicoccum* wheats by interspecific hybridization with 4N *Triticum* species. Indian J Genet Plant Breed 39:114–125

de Pauw RM (1978) Breeding for post-seedling resistance to wheat stem rust. Cereal Res Commun 6:249–253

Dewey DR (1984) The genome system of classification as a guide to intergeneric hybridization with the perennial Triticeae. In: Gustafson JP (ed) Gene manipulation in plant improvement. 16th Stadler Genetics Symp, Plenum, New York, pp 209–279

Dhaliwal AS, Mares DJ, Marshall DR (1987) Effect of 1B/1R chromosome translocation on milling and quality characteristics of bread wheats. Cereal Chem 64:72–76

Driscoll CJ, Jensen NF (1963) A genetic method for detecting induced intergeneric translocations. Genetics 48:459–468

Driscoll CJ, Jensen NF (1964) Characteristics of leaf rust resistance transferred from rye to wheat. Crop Sci 4:372–374

Driscoll CJ, Gordon GH, Kimber G (1980) Mathematics of chromosome pairing. Genetics 95:159–169

Dvořák J (1977) Transfer of leaf rust resistance from *Aegilops speltoides* to *Triticum aestivum*. Can J Genet Cytol 19:133–141

Dyck PL (1977) Genetics of leaf rust reaction in three introductions of common wheat. Can J Genet Cytol 19:711–716

Dyck PL (1979) Identification of the gene for adult-plant leaf rust resistance in Thatcher. Can J Plant Sci 59:499–501

Dyck PL (1987) The association of a gene for leaf rust resistance with the chromosome 7D suppressor of stem rust resistance in common wheat. Can J Genet Cytol 29:467–469

Dyck PL, Johnson R (1983) Temperature sensitivity of genes for resistance in wheat to *Puccinia recondita*. Can J Plant Pathol 5:229–234

Dyck PL, Kerber ER (1970) Inheritance in hexaploid wheat of adult-plant leaf rust resistance derived from *Aeqilops squarrosa*. Can J Genet Cytol 12:175–180

Dyck PL, Kerber ER (1981) Aneuploid analysis of a gene for leaf rust resistance derived from the common wheat cultivar Terenzio. Can J Genet Cytol 23:405–409

Dyck PL, Samborski DJ (1968a) Genetics of resistance of leaf rust in the common wheat varieties Webster, Loros, Brevit, Carina, Malakof and Centenario. Can J Genet Cytol 10:7–17

Dyck PL, Samborski DJ (1968b) Host-parasite interactions involving two genes for leaf rust resistance in wheat. In: Finlay KW, Shepherd KW (eds) Proc 3rd Int Wheat Genetics Symp, Canberra, pp 245–250

Dyck PL, Samborski DJ (1970) The genetics of two alleles for leaf rust resistance at the *Lr14* locus in wheat. Can J Genet Cytol 12:689–694

Dyck PL, Samborski DJ (1974) Inheritance of virulence in *Puccinia recondita* on alleles at the *Lr2* locus for resistance in wheat. Can J Genet Cytol 16:323–332

Dyck PL, Samborski DJ (1982) The inheritance of resistance to *Puccinia recondita* in a group of common wheat cultivars. Can J Genet Cytol 24:273–283

Dyck PL, Samborski DJ, Anderson RG (1966) Inheritance of the adult plant leaf rust resistance derived from the common wheat varieties Exchange and Frontana. Can J Genet Cytol 8:665–671

Dyck PL, Kerber ER, Lukow OM (1987) Chromosome location and linkage of a new gene (*Lr33*) for reaction to *Puccinia recondita* in common wheat. Genome 29:463–466

Ellingboe AH (1961) Somatic recombination in *Puccinia graminis* var. *tritici*. Phytopathology 51:13–15

Elliot FC (1958) Plant breeding and cytogenetics. McGraw-Hill, New York

Endo TR (1982) Gametocidal chromosomes of the *Aegilops* species in common wheat. Can J Genet Cytol 24:201–206

Eyal Z, Yurman R, Moseman JG, Wahl I (1973) Use of mobile nurseries in pathogenicity studies of *Puccinia graminis hordei* on *Hordeum spontaneum*. Phytopathology 63:1330–1334

Falk DE, Kasha KJ (1981) Comparison of crossability of rye (*Secale cereale*) and *Hordeum bulbosum* onto wheat (*Triticum aestivum*). Can J Genet Cytol 23:81–88

Fedak G (1980) Production, morphology and meiosis of reciprocal barley-wheat hybrids. Can J Genet Cytol 22:117–123

Fitzgerald PJ, Caldwell RM, Nelson OE (1957) Inheritance of resistance to certain races of leaf rust of wheat. Agron J 49:539–543

Fleischmann G, Baker RJ (1971) Oat crown rust race differentiation: replacement of the standard differential varieties with a new set of single resistance gene lines derived from *Avena sterilis*. Can J Bot 49:1433–1437

Flor HH (1942) Inheritance of pathogenicity in *Melampsora lini*. Phytopathology 32:653–669

Flor HH (1946) Genetics of pathogenicity in *Melampsora lini*. J Agric Res 73:335–357

Flor HH (1947) Inheritance of reaction to rust in flax. J Agric Sci 74:241–262

Flor HH (1956) The complementary genic systems in flax and flax rust. Adv Genet 8:29–54

Forsyth FR (1956) Interaction of temperature and light on the seedling reaction of McMurachy wheat to race 15B of stem rust. Can J Bot 34:745–749

Frey KJ, Browning JA, Simons MD (1973) Management of host resistance genes to control diseases. Z Pflanzenkrankh Pflanzenschutz 80:160–180

Frey KJ, Browning JA, Simons MD (1979) Management systems for host genes to control disease loss. Ind J Genet Plant Breed 39:10–21

Fuchs E (1960) Physiologische Rassen bei Gelbrost *Puccinia glumarum* (Schm.) Erikss. et Henn. auf weizen. Nachr Dtsch Pflanzenschutzbd Braunschweig 12:49–63

Gaines R (1976) The genetic analysis and description of durable resistance to yellow rust of wheat. D Phil Thesis, Univ Cambridge

Gamborg OL, Shyluk JP (1981) Nutrition, media and characteristics of plant cell and tissue cultures. In: Thorpe TA (ed) Plant tissue culture: methods and application in agriculture. Academic, New York London, pp 21–44

Gassner G, Straib W (1932) Die Bestimmung der biologischen Rassen des Weizengelbrostes (*Puccinia glumarum* f. sp. *tritici* (Schmidt) Erikss. and Henn.) Arb Biol Reichsanst Land Fortswirtsch 20:141–163

Gerechter-Amitai ZK, Grama A (1974) Inheritance of resistance to stripe rust (*Puccinia striiformis*) in crosses between wild emmer (*Triticum dicoccoides*) and cultivated tetraploid and hexaploid wheats. I. *Triticum durum*. Euphytica 23:387–392

Gerechter-Amitai ZK, Wahl I, Vardi A, Zohary D (1971) Transfer of stem rust seedling resistance from wild diploid einkorn to tetraploid *durum* wheat by means of a triploid hybrid bridge. Euphytica 20:281–285

Gill BS, Browder LE, Hatchett JH, Harvey TL, Martin TJ, Raupp WJ, Sharma HC, Waines JG (1983) Disease and insect resistance in wild wheat. In: Sakamoto S (ed) Proc 6th Int Wheat genetics Symp, Kyoto, pp 785–792

Gill KS, Nanda GS, Singh G, Aujla SS (1980) Studies on multilines in wheat (*Triticum aestivum* L). 12. Breeding of multiline variety by convergence of breeding lines. Euphytica 29:125–128

Gilmour J (1973) Octal notation for designating physiologic races of plant pathogens. Nature (London) 242:620

Goddard MV (1976a) Cytological studies of *Puccinia striiformis* (Yellow rust of wheat). Trans Br Mycol Soc 66:433–437

Goddard MV (1976b) The production of a new race, 105 E137 of *Puccinia striiformis* in glasshouse experiments. Trans Br Mycol Soc 67:395–398

Gough FJ, Merkle OG (1971) Inheritance of stem and leaf rust resistance in Agent and Agrus cultivars of *Triticum aestivum*. Phytopathology 61:1501–1505

Grafius JE (1965) Short cuts in plant breeding. Crop Sci 5:377

Grama A, Gerechter-Amitai ZK (1974) Inheritance of resistance to stripe rust (*Puccinia striiformis*) in crosses between wild emmer (*Triticum dicoccoides*) and cultivated tetraploid and hexaploid wheats. II. *Triticum aestivum*. Euphytica 23:393–398

Grama A, Gerechter-Amitai ZK, Blum A (1983) Wild emmer as a donor of genes for resistance to stripe rust and for high protein content. In: Sakamoto S (ed) Proc 6th Int Wheat genetics Symp, Kyoto, pp 187–192

Green GJ (1964) A color mutation, its inheritance, and the inheritance of pathogenicity in *Puccinia graminis* Pers. Can J Bot 42:1653–1664

Green GJ (1966) Selfing studies with races 10 and 11 of wheat stem rust. Can J Bot 44:1255–1260

Green GJ (1975) Virulence changes in *Puccinia graminis* f.sp. *tritici* in Canada. Can J Bot 53:1377–1386

Green GJ (1981) Identification of physiologic races of *Puccinia graminis* f.sp. *tritici* in Canada. Can J Plant Pathol 3:33–39

Green GJ, Campbell AB (1979) Wheat cultivars resistant to *Puccinia graminis tritici* in western Canada: their development, performance, and economic value. Can J Plant Pathol 1:3–11

Green GJ, Knott DR, Watson IA, Pugsley AT (1960) Seedling reactions to stem rust of lines of wheat with substituted genes for rust resistance. Can J Plant Sci 40:524–538

Groenewegen LJM (1977) Multilines as a tool in breeding for reliable yields. Cereal Res Commun 5:125–132

Groenewegen LJM, Zakoks JC (1979) Exploiting within-field diversity as a defence against cereal diseases: a plea for poly-genotype varieties. Indian J Genet Plant Breed 39:81–94

Habgood RM (1970) Designation of physiological races of plant pathogens. Nature (London) 227:1268–1269

Haggag MEA, Dyck PL (1973) The inheritance of leaf rust resistance in four common wheat varieties possessing genes at or near the *Lr3* locus. Can J Genet Cytol 15:127–134

Haggag MEA, Samborski DJ, Dyck PL (1973) Genetics of pathogenicity in three races of leaf rust on four wheat varieties. Can J Genet Cytol 15:73–82

Hanchinal RR, Good JV (1982) Seed setting and germination of AB-genome monosomic lines of PbC591 × Bijaga Yellow and their backcross generations in wheat. Wheat Inf Serv 55:15–21

Harder DE (1984) Developmental ultrastructure of hyphae and spores. In: Bushnell WR, Roelfs AP (eds) The cereal rusts, vol 1. Origins, specificity, structure and physiology. Academic Press, New York London Orlando, pp 333–373

Harder DE, Chong J (1984) Structure and physiology of haustoria. In: Bushell WR, Roelfs AP. The cereal rusts, vol 1. Origins, specificity, structure, and physiology. Academic Press, New York London Orlando, pp 431–476

Hayes HK, Parker JH, Kurtzweil C (1920) Genetics of rust resistance in crosses of varieties of *Triticum vulgare* with varieties of *T. durum* and *T. dicoccum*. J Agric Res 19:523–542

Hobbs CD, Merkle OG (1965) Improved technique using calcium pantothenate for hypodermic inoculation of wheat with *Puccinia graminis* f.sp. *tritici*. Crop Sci 5:192

Hungerford CW, Owens CE (1923) Specialized varieties of *Puccinia glumarum* and hosts for variety *tritici*. J Agric Sci 25:363–401

Jan C-C, Dvorak J, Qualset CO, Soliman KM (1981) Selection and identification of a spontaneous alien chromosome translocation in wheat. Genetics 98:389–398

Jensen CJ (1974) Chromosome doubling techniques in haploids. In: Kasha KJ (ed) Haploids in higher plants – advances and potentials. Univ Press, Guelph, pp 153–190

Jensen NF (1952) Intra-varietal diversification in oat breeding. Agron J 44:30–34

Johnson R (1978) Induced resistance to fungal diseases with special reference to yellow rust of wheat. Ann Appl Biol 89:107–110

Johnson R (1979) The concept of durable resistance. Phytopathology 69:198–199

Johnson R (1980) Genetics of adult plant resistance to yellow rust in winter wheat cultivars. Proc 5th European and Mediterranean cereal rusts Conf, Bari Rome, pp 59–63

Johnson R (1981) Durable resistance: definition of, genetic control, and attainment in plant breeding. Phytopathology 71:567–568

Johnson R (1983) Genetic background of durable resistance. In: Lamberti F, Waller JM, Van der Graaff NA (eds) Durable resistance in crops. Plenum, New York, pp 5–26

Johnson R (1984) A critical analysis of durable resistance. Annu Rev Phytopathol 22:309–330

Johnson R, Dyck PL (1984) Resistance to yellow rust in *Triticum spelta* var. *album* and bread wheat cultivars Thatcher and Lee. Proc 6th European and Mediterranean cereal rusts Conf, Grignon. Coll INRA 25:71–74

Johnson R, Lupton FGH (1987) Breeding for disease resistance. In: Lupton FGH (ed) Wheat breeding: its scientific basis. Chapman & Hall, London New York, pp 369–424

Johnson R, Taylor AJ (1976) Effects of resistance induced by non-virulent races of *Puccinia striiformis*. In: Brönnimann A (ed) 4th European and Mediterranean cereal rusts Conf, Interlaken, pp 49–51

Johnson R, Stubbs RW, Fuchs E, Chamberlain NH (1972)Nomenclature for physiologic races of *Puccinia striiformis* infecting wheat. Trans Br Mycol Soc 58:475–480

Johnson R, Taylor AJ, Smith GMB (1986) Yellow rust of wheat. Plant Breed Inst Annu Rep 1985:87–88

Johnson R, Stubbs RW, Kirmani MAS, Rizvi SSA, Statler GD (1987) Discussion of a method resulting in erroneous postulation of the gene *Yr8* for resistance to *Puccinia striiformis* in Pakistani wheat cultivars. Cereal Rusts Bull 15:13–19

Johnson T (1949) Intervarietal crosses in *Puccinia graminis*. Can J Res Sect C 27:45–65

Johnson T (1961) Man-guided evolution in plant rusts. Science 133:357–362

Johnston CO, Heyne EG (1964) Wichita wheat backcross lines for differential hosts in identifying physiologic races of *Puccinia recondita*. Phytopathology 54:385–388

Johnston CO, Mains EB (1932) Studies on physiologic specialization in *Puccinia triticina*. US Dep Agric Tech Bull 313, 22 pp

Joppa LR (1987) Aneuploid analysis in tetraploid wheat. In: Heyne EG (ed) Wheat and wheat improvement. Am Soc Agron, Madison, pp 255–267

Joppa LR, Williams ND (1977) D-genome substitution-monosomics of durum wheat. Crop Sci 17:772–776

Joppa LR, Williams ND (1988) Langdon durum disomic substitution lines and aneuploid analysis in tetraploid wheat. Genome 30:222–228

Joshi BL, Singh D (1979) Introduction of alien variation into bread wheat. In: Ramanujam S (ed) Proc 5th Intern Wheat genetics Symp, New Delhi, pp 342–348

Kaltsikes PJ (1974) Methods for Triticale production. Z Pflanzenzücht 71:264–286

Kao KN, Knott DR (1969) The inheritance of pathogenicity in races 111 and 29 of wheat stem rust. Can J Genet Cytol 11:266–274

Kerber ER (1983) Suppression of rust resistance in amphiploids of *Triticum*. In: Sakamoto S (ed) Proc 6th Int Wheat genetics Symp, Kyoto, pp 813–817

Kerber ER, Dyck PL (1969) Inheritance in hexaploid wheat of leaf rust resistance and other characters derived from *Aegilops squarrosa*. Can J Genet Cytol 11:639–647

Kerber ER, Dyck PL (1973) Inheritance of stem rust resistance transferred from diploid wheat (*Triticum monococcum*) to tetraploid and hexaploid wheat and chromosome location of the gene involved. Can J Genet Cytol 15:397–409

Kerber ER, Dyck PL (1979) Resistance to stem and leaf rust of wheat in *Aegilops squarrosa* and transfer of a gene for stem rust resistance to hexaploid wheat. In: Ramanujam S (ed) Proc 5th Int Wheat genetics Symp, New Delhi, pp 358–364

Kerber ER, Green GJ (1980) Suppression of stem rust resistance in hexaploid wheat cv. Canthatch by chromosome 7DL. Can J Bot 58:1347–1350

Kerby K, Kuspira J (1987) The phylogeny of the polyploid wheats *Triticum aestivum* (bread wheat) and *Triticum turgidum* (macaroni wheat). Genome 29:722–737

Kibirige-Sebunya I, Knott DR (1983) Transfer of stem rust resistance to wheat from an *Agropyron* chromosome having a gametocidal effect. Can J Genet Cytol 25:215–221

Kihara H (1919) Über cytologische Studien bei einigen Getreidearten. I. Species-Bastarde des Weizens und Weizenroggen-Bastarde. Bot Mag (Tokyo) 33:17–38

Kihara H (1924) Cytologische und genetische Studien bei wichtigen Getreidearten mit besonderer Rücksicht auf das Verhalten der Chromosomen und die Sterilität in den Bastarden. Mem Coll Sci, Kyoto Imp Univ, Ser B:1–200

Kimber G (1983) Genome analysis in the genus Triticum. In: Sakamoto S (ed) Proc 6th Int Wheat genetics Symp, Kyoto, pp 23–28

Kimber G, Sears ER (1987) Evolution in the genus *Triticum* and the origin of cultivated wheat. In: Heyne EG (ed) Wheat and wheat improvement. Am Soc Agron, Madison, pp 154–164

Kirmani MAS, Rizvi SS, Stubbs RW (1984) Postulated genotypes for stripe rust resistance in wheat cultivars of Pakistan. Proc 6th European Mediterranean cereal rusts Conf, Grignon, pp 81–85

Knoblock IW (1968) A check list of crosses in the Gramineae. Michigan State Univ Press, East Lansing, 170 pp

Knott DR (1957) The inheritance of rust resistance. II. The inheritance of stem rust resistance in six additional varieties of common wheat. Can J Plant Sci 37:177–192

Knott DR (1959) The inheritance of rust resistance. IV. Monosomic analysis of rust resistance and some other characters in six varieties of wheat including Gabo and Kenya Farmer. Can J Plant Sci 39:215–228

Knott DR (1961) The inheritance of rust resistance. VI. The transfer of stem rust resistance from *Agropyron elongatum* to common wheat. Can J Plant Sci 41:109–123

Knott DR (1962a) Inheritance of rust resistance. VIII. Additional studies on Kenya varieties of wheat. Crop Sci 2:130–132

Knott DR (1962b) The inheritance of rust resistance. IX. The inheritance of resistance to races 15B and 56 of stem rust in the wheat variety Khapstein. Can J Plant Sci 42:415–419

Knott DR (1963) Note on Stewart 63 durum wheat. Can J Plant Sci 43:605–607

Knott DR (1968) The inheritance of resistance to stem rust races 56 and 15B-1L(Can.) in the wheat varieties Hope and H-44. Can J Genet Cytol 10:311–320

Knott DR (1971a) Genes for stem rust resistance in Hope and H-44. Can J Genet Cytol 13:186–188

Knott DR (1971b) The transfer of genes for disease resistance from alien species to wheat by induced translocations. In: IAEA (ed) Mutation breeding for disease resistance. IAEA, Vienna, pp 67–77

Knott DR (1980) Mutation of a gene for yellow pigment linked to *Lr19* in wheat. Can J Genet Cytol 22:651–654

Knott DR (1981) The effects of genotype and temperature on the resistance to *Puccinia graminis tritici* controlled by the gene *Sr6* in *Triticum aestivum*. Can J Genet Cytol 23:183–190

Knott DR (1982) Multigenic inheritance of stem rust resistance in wheat. Crop Sci 22:393–399

Knott DR (1984) The association and dissociation of genes for virulence in wheat stem rust. Phytopathology 74:1023

Knott DR (1986a) The field reaction to stem rust of 'Chinese Spring' substitution lines carrying chromosomes from 'Hope' and 'Thatcher' wheats. Can J Genet Cytol 28:12–16

Knott DR (1986b) Novel approaches to wheat breeding. In: Smith EL (ed) Genetic improvement in yield of wheat. Am Soc Agron Spec Publ 13:25–40

Knott DR (1987) Transferring alien genes to wheat. In: Heyne EG (ed) Wheat and wheat improvement, 2nd edn. Am Soc Agro, Madison, pp 462–471

Knott DR, Anderson RG (1956) The inheritance of rust resistance. I. The inheritance of stem rust resistance in ten varieties of common wheat. Can J Agric Sci 36:174–195

Knott DR, Green GJ (1964) Seedling reactions to stem rust of lines of Marquis wheat with substituted genes for rust resistance. Can J Plant Sci 45:106–107

Knott Dr, McIntosh RA (1978) The inheritance of stem rust resistance in the common wheat cultivar 'Webster'. Crop Sci 18:365–369

Knott DR, Padidam M (1988) Inheritance of resistance to stem rust in six wheat lines having adult plant resistance. Genome 30:283–288

Konzak CF, Joppa LR (1988) The inheritance and chromosomal location of a gene for chocolate chaff in durum wheat. Genome 30:229–230

Krupinsky JM, Sharp EL (1978) Additive resistance in wheat to *Puccinia striiformis*. Phytopathology 68:1795–1799

Krupinsky JM, Sharp EL (1979) Reselection for improved resistance of wheat to stripe rust. Phytopathology 69:400–404

Kuhn RC, Ohm HW, Shaner G (1980) Inheritance of slow leaf-rusting resistance in Suwon 85 wheat. Crop Sci 20:655–659

Kuspira J (1966) Intervarietal chromosome substitution in hexaploid wheat. In: Mac Key J (ed) Proc 2nd Int Wheat genetics Symp, Lund. Hereditas Suppl 2:355–369

Kuspira J, Unrau J (1957) Genetic analysis of certain characters in common wheat using whole chromosome substitution lines. Can J Plant Sci 37:300–326

Lamberti F, Waller JM, Van der Graaff NA (eds) (1983) Durable resistance in crop plants. Plenum, New York

Lapitan NLV, Sears RG, Gill BS (1984) Translocations and other karyotypic structural changes in wheat × rye hybrids regenerated from tissue culture. Theor Appl Genet 68:547–554

Larkin PJ, Scowcroft WR (1981) Somaclonal variation – a novel source of variability from cell cultures for plant improvement. Theor Appl Genet 60:197–214

Larkin PJ, Ryan SA, Brettell RIS, Scowcroft WR (1984) Heritable somaclonal variation in wheat. Theor Appl Genet 67:443–455

Law CN (1967) The location of genetic factors controlling a number of quantitative characters in wheat. Genetics 56:445–461

Law CN, Johnson R (1967) A genetic study of leaf rust resistance in wheat. Can J Genet Cytol 9:805–822

Law CN, Wolfe MS (1966) Location of genetic factors for mildew resistance and ear emergence time on chromosome 7B of wheat. Can J Genet Cytol 8:462–470

Law CN, Snape JW, Worland AJ (1987) Aneuploidy in wheat and its uses in genetic analysis. In: Lupton FGH (ed) Wheat breeding: its scientific basis. Chapman & Hall, London New York, pp 71–107

Lein A (1943) Die genetische Grundlage der Kreuzbarkeit zwischen Weizen and Roggen. Z Indukt Abstammungs Vererbungsl 81:28–61

Leppik EE (1967) Some viewpoints on the phylogeny of rust fungi. VI. Biogenic radiation. Cytologia 59:568–579

Leppik EE (1970) Gene centres of plants as sources of disease resistance. Annu Rev Phytopathol 8:324–344

Lewellen RT, Sharp EL (1968) Inheritance of minor gene combinations in wheat to *Puccinia striiformis* at two temperature profiles. Can J Bot 46:21–26

Lewellen RT, Sharp EL, Hehn ER (1967) Major and minor genes in wheat for resistance to *Puccinia striiformis* and their responses to temperature changes. Can J Bot 45:2155–2172

Lilienfeld FA (1951) H. Kihara: Genome-analysis in *Triticum* and *Aegilops*. Concluding review. Cytologia 16:101–123

Line RF, Sharp EL, Powelson RL (1970) A system for differentiating races of *Puccinia striiformis* in the United States. Plant Dis Rep 54:992–994

Little R, Manners JG (1969) Somatic recombination in yellow rust of wheat, *Puccinia striiformis*. I. The production and possible origin of two new physiologic races. Trans Br Mycol Soc 53:251–258

Littlefield LJ (1981) Biology of the plant rusts. Iowa State Univ Press, Ames

Loegering WQ (1966) The relationship between host and pathogen in stem rust of wheat. In: Mac Key J (ed) Proc 2nd Int Wheat Genetics Symp, Lund. Hereditas Suppl 2:167–177

Loegering WQ (1984) Genetics of the pathogen-host association. In: Bushnell WR, Roelfs AP (eds) The cereal rusts, vol 1. Origins, specificity, structure and physiology. Academic Press, New York London Orlando, pp 165–192

Loegering WQ, Burton CH (1974) Computer-generated hypothetical genotypes for reaction and pathogenicity of wheat cultivars and cultures of *Puccinia graminis tritici*. Phytopathology 64:1380–1384

Loegering WQ, Geis JR (1957) Independence in the action of three genes conditioning stem rust resistance in Red Egyptian wheat. Phytopathology 47:740–741

Loegering WQ, Powers HR (1962) Inheritance of pathogenicity in a cross of physiological races 111 and 36 of *Puccinia graminis* f.sp. *tritici*. Phytopathology 52:547–554

Loegering WQ, Sears ER (1963) Distorted inheritance of stem-rust resistance of Timstein wheat caused by a pollen-killing gene. Can J Genet Cytol 5:65–72

Loegering WQ, Sears ER (1966) Relationships among stem-rust genes on wheat chromosomes 2B, 4B and 6B. Crop Sci 6:157–160

Loegering WQ, Johnson CO, Samborski DJ, Caldwell RM, Schafer JF, Young HC, Jr. (1959) A proposed modification of the system of wheat leaf rust race identification and nomenclature. Plant Dis Rep 43:613–615

Loegering WQ, Johnson CO, Samborski DJ, Caldwell RM, Schafer JF, Young HC, Jr. (1961a) The North American 1961 set of supplemental differential wheat varieties for leaf rust race identification. Plant Dis Rep 45:444–447

Loegering WQ, McKinney HH, Harmon DL, Clark WA (1961b) A long term experiment for preservation of urediospores of *Puccinia graminis tritici* in liquid nitrogen. Plant Dis Rep 45:384–385

Loegering WQ, McIntosh RA, Burton CH (1971) Computer analysis of disease data to derive hypothetical genotypes for reaction of host varieties to pathogens. Can J Genet Cytol 13:742–748

Löffler CM, Busch RH, Wiersma JV (1983) Recurrent selection for grain protein percentage in hard red spring wheat. Crop Sci 23:1097–1101

Löve A (1982) Generic evolution of the wheatgrasses. Biol Zentralbl 101:199–212

Luig NH (1968) Mechanisms of differential transmission of gametes in wheat. In: Finlay KW, Shepherd KW (eds) Proc 3rd Int Wheat genetics Symp, Aust Acad Sci, Canberra, pp 322–323

Luig NH (1979) Mutation studies in *Puccinia graminis tritici*. In: Ramanujam (ed) Proc 5th Int Wheat genetics Symp, New Delhi, pp 533–539

Luig NH (1983) A survey of virulence genes in wheat stem rust, *Puccinia graminis* f.sp. *tritici*. Parey, Berlin Hamburg, 199 pp

Luig NH (1985) Epidemiology in Australia and New Zealand. In: Roelfs AP, Bushnell WR (eds) The cereal rusts, vol 2. Diseases, distribution, epidemiology, and control. Academic Press, New York London Orlando, pp 301–328

Luig NH, McIntosh RA (1968) Location and linkage of genes on wheat chromosome 2D. Can J Genet Cytol 10:99–105

Luig NH, Rajaram S (1972) The effect of temperature and genetic background on host gene expression and interaction to *Puccinia graminis tritici*. Phytopathology 61:1171–1174

Luig NH, Watson IA (1961) A study of inheritance of pathogenicity in *Puccinia graminis* var. *tritici*. Proc Linn Soc NSW 86:217–229

Luig NH, Watson IA (1967) Vernstein – a *Triticum aestivum* derivative with Vernal emmer-type stem rust resistance. Crop Sci 7:31–33

Lupton FGH, Johnson R (1970) Breeding for mature-plant resistance to yellow rust in wheat. Ann Appl Biol 66:137–143

Lupton FGH, Macer RCF (1962) Inheritance of resistance to yellow rust (*Puccinia glumarum* Erikss & Henn) in seven varieties of wheat. Trans Br Mycol Soc 45:21–45

Luthra JK, Rao MV (1979) Escape mechanism operating in multilines and its significance in relation to leaf rust epidemics. Indian J Genet Plant Breed 39:38–49

Macer RCF (1975) Plant pathology in a changing world. Trans Br Mycol Soc 65:351–367

Maclean DJ (1982) Axenic culture and metabolism of rust fungi. In: Scott KJ, Chakravorty AK (eds) The rust fungi. Academic Press, London New York, pp 37–84

Maddock SE, Semple JT (1986) Field assessment of somaclonal variation in wheat. J Exp Bot 37:1065–1078

Mains EB, Jackson HS (1926) Physiologic specialization in the leaf rust of wheat *Puccinia triticina* Erikss. Phytopathology 16:89–120

Mains EB, Leighty CE, Johnston CO (1926) Inheritance of resistance to leaf rust *Puccinia triticina* Erikss., in crosses of common wheat, *Triticum vulgare* Vill. J Agric Res 32:931–972

Manners JG (1950) Studies on the physiologic specialization of yellow rust (*Puccinia glumarum* (Schm.) Erikss and Henn) in Great Britain. Ann Appl Biol 37:187–214

Marshall DR, Burdon JJ, Muller WJ (1986) Multiline varieties and disease control. 6. Effects of selection at different stages of the pathogen life cycle on the evolution of virulence. Theor Appl Genet 71:801–809

Martin CD, Littlefield LJ, Miller JD (1977) Development of *Puccinia graminis* f.sp. *tritici* in seedling plants of slow rusting wheats. Trans Br Mycol Soc 68:161–166

Martin CD, Miller JD, Busch RH, Littlefield LJ (1979) Quantitation of slow rusting in seedling and adult spring wheat. Can J Bot 57:1550–1556

Martinez-Gonzalez JMS, Wilcoxson RD, Stuthman DD, McVey DV, Busch RH (1983) Genetic factors conditioning slow rusting in Era wheat. Phytopathology 73:247–249

McEwan JM, Kaltsikes PJ (1970) Early generation testing as a means of predicting the value of specific chromosome substitutions into common wheat. Can J Genet Cytol 12:711–723

McFadden ES (1930) A successful transfer of emmer characters to *vulgare* wheat. J Am Soc Agron 22:1020–1034

McGinnis RC (1953) Cytological studies of chromosomes of rust fungi. I. The mitotic chromosomes of *Puccinia graminis*. Can J Bot 31:522–526

McGinnis RC (1956) Cytological studies of chromosomes of rust fungi. J Hered 47:255–259

McIntosh RA (1973) A catalogue of gene symbols for wheat. In: Sears ER, Sears LMS (eds) Proc 4th Int Wheat genetics Symp, Columbia, pp 893–937

McIntosh RA (1978) Cytogenetical studies in wheat. X. Monosomic analysis and linkage studies involving genes for resistance to *Puccinia graminis* f.sp. *tritici* in cultivar Kota. Heredity 41:71–82

McIntosh RA (1983) A catalogue of gene symbols for wheat (1983 edn). In: Sakamoto S (ed) Proc 6th Int Wheat Genetics Symp, Kyoto, pp 1197–1254

McIntosh RA (1986) Catalogue of gene symbols for wheat: 1986 supplement. Cereal Res Commun 14:105–115

McIntosh RA (1987a) Catalogue of gene symbols for wheat: 1987 supplement. Ann Wheat Newslett 33;196

McIntosh RA (1987b) Gene location and gene mapping in hexaploid wheat. In: Heyne EG (ed) Wheat and wheat improvement. Am Soc Agron, Madison, pp 269–287

McIntosh RA, Baker EP (1968) A linkage map for chromosome 2D. In: Finlay KW, Shepherd KW (eds) 3rd Int Wheat Genetics Symp, Canberra, pp 305–309

McIntosh RA, Cusick JE (1987) Linkage map of hexaploid wheat. In: Heyne EG (ed) Wheat and wheat improvement. Am Soc Agron, Madison, pp 289–297

McIntosh RA, Dyck PL (1975) Cytogenetical studies in wheat. VII. Gene *Lr23* for reaction to *Puccinia recondita* in Gabo and related cultivars. Aust J Biol Sci 28:201–211

McIntosh RA, Gyarfas J (1971) *Triticum timopheevii* as a source of resistance to wheat stem rust. Z Pflanzenzücht 66:240–248

McIntosh RA, Luig NH (1973) Linkage of genes for reaction to *Puccinia graminis* f.sp. *tritici* and *P recondita* in Selkirk wheat and related cultivars. Aust J Biol Sci 26:1145–1152

McIntosh RA, Watson IA (1982) Genetics of host pathogen interactions in rusts. In: Scott AK, Chakravorty AK (eds) The rust fungi. Academic Press, New York London, pp 121–149

McIntosh RA, Luig NH, Baker EP (1967) Genetic and cytogenetic studies of stem rust, leaf rust and powdery mildew resistances in Hope and related wheat cultivars. Aust J Biol Sci 20:1181–1192

McIntosh RA, Dyck PL, Green GJ (1974) Inheritance of reaction to stem rust and leaf rust in the wheat cultivar Etoile de Choisy. Can J Genet Cytol 16:571–577

McIntosh RA, Dyck PL, Green GJ (1977) Inheritance of leaf and stem rust resistances in wheat cultivars Agent and Agatha. Aust J Agric Res 28:37–45

McIntosh RA, Partridge M, Hare RA (1980) Telocentric mapping of *Sr12* in wheat chromosome 3B. Cereal Res Commun 8:321–324

McIntosh RA, Miller TE, Chapman V (1982) Cytogenetical studies in wheat. XII. *Lr28* for resistance to *Puccinia recondita* and *Sr34* for resistance to *P. graminis tritici*. Z Pflanzenzücht 89:295–306

McIntosh RA, Luig NH, Milne DL, Cusick J (1983) Vulnerability of triticales to wheat stem rust. Can J Plant Pathol 5:61–69

McIntosh RA, Dyck PL, The TT, Cusick J, Milne DL (1984) Cytogenetical studies in wheat. XIII. *Sr35* — a third gene from *Triticum monococcum* for resistance to *Puccinia graminis tritici*. Z Pflanzenzücht 92:1–14

McIntosh RA, Hart OE, Gale MD (1987) Catalogue of gene symbols for wheat: 1987 supplement. Ann Wheat Newsl 33:191–202

McNeal FH, McGuire CF, Berg MA (1978) Recurrent selection for grain protein content in spring wheat. Crop Sci 18:779–782

Mendgen K (1984) Development and physiology of teliospores. In: Roelfs AP, Bushnell WR (eds) The cereal rusts, vol 1. Origin, specificity, structure and physiology. Academic Press, New York London Orlando, pp 375–398

Merkle OG, Starks KJ (1985) Resistance of wheat to the yellow sugarcane aphid (Homoptera: Aphidae). J Econ Entomol 78:127–128

Mettin D, Bluthner WD, Schlegel G (1973) Additional evidence on spontaneous IB/IR wheat-rye substitutions and translocations. In: Sears ER, Sears LMS (eds) Proc 4th Int Wheat genetics Symp, Columbia, pp 179–184

Miller TE (1987) Systematics and evolution. In: Lupton FGH (ed) Wheat breeding: its scientific basis. Chapman & Hall, London New York, pp 1–30

Miller TE, Hutchinson J, Chapman V (1982) Investigation of a preferentially transmitted *Aegilops sharonensis* chromosome in wheat. Theor Appl Genet 61:27–33

Mochizuki A (1968) The monosomics of durum wheat In: Finlay KW, Shepherd KW (eds) Proc 3rd Int Wheat genetics Symp, Canberra, pp 310–315

Modawi RS, Browder LE, Heyne EG (1985) Use of infection-type data to identify genes for low reaction to *Puccinia recondita* in several winter wheat cultivars. Crop Sci 25:9–13

Morris R, Sears ER (1967) The cytogenetics of wheat and its relatives. In: Quisenberry KS, Reitz LP (eds) Wheat and wheat improvement. Am Soc Agron, Madison, pp 19–87

Mundt CC, Browning JA (1985) Genetic diversity and cereal rust management. In: Roelfs AP, Bushnell WR (eds) The cereal rusts, vol 2. Diseases, distribution, epidemiology and control. Academic Press, New York London Orlando, pp 527–560

Murashige T, Skoog F (1962) A revised medium for rapid growth and bioassays with tobacco tissue cultures. Physiol Plant 15:473–497

Murphy JP, Helsel DB, Elliott A, Thro AM, Frey KJ (1982) Compositional stability of an oat multiline. Euphytica 31:33–40

Nelson RR (1956) Transmission of factors for urediospore color in *Puccinia graminis* var. *tritici* by means of nuclear exchange between vegetative hyphae. Phytopathology 46:538–540

Nelson RR, Wilcoxson RD, Christensen JJ (1955) Heterocaryosis as a basis for variation in *Puccinia graminis tritici*. Phytopathology 45:639–643

Newton AC, Caten CE, Johnson R (1985) Variation for isozymes and double-stranded RNA among isolates of *Puccinia striiformis* and two other cereal rusts. Plant Pathol 34:235–247

Newton M, Johnson T, Brown AM (1930a) A preliminary study on the hybridization of physiologic forms of *Puccinia graminis tritici*. Sci Agr 10: 721–731

Newton M, Johnson T, Brown AM (1930b) A study of the inheritance of spore color and pathogenicity in crosses between physiologic forms of *Puccinia graminis tritici*. Sci Agr 10:775–798

Nyquist WE (1962) Differential fertilization in the inheritance of stem rust resistance in hybrids involving a common wheat strain derived from *T. timopheevii*. Genetics 47:1109–1124

Ohm HW, Shaner GE (1976) Three components of slow leaf-rusting at different growth stages in wheat. Phytopathology 66:1356–1360

Okamoto M (1957) Asynaptic effect of chromosome V. Wheat Inf Serv 5:6

Okamoto M (1962) Identification of the chromosomes of common wheat belonging to the A and B genomes. Can J Genet Cytol 14:31–37

Okamoto M (1966) Studies on chromosome 5B effects in wheat. In: Mac Key J (ed) Proc 2nd Int Wheat genetics Symp, Lund. Hereditas Suppl 2:409–417

Pandey HN, Rao MV (1983) Role of temperature in restoration of fertility in *Triticum durum* × *T. timopheevi* hybrids. Z Pflanzenzücht 91:70–73

Parlevliet JE (1977) Evidence of differential interaction in the polygenic *Hordeum vulgare-Puccinia hordei* relation during epidemic development. Phytopathology 67:776–778

Parlevliet JE (1979) Components of resistance that reduce the rate of epidemic development. Annu Rev Phytopathol 17:203–222

Parlevliet JE (1983) Models explaining the specificity and durability of host resistance derived from the observations on the barley-*Puccinia hordei* system. In: Lamberti F, Waller JM, Van der Graaff NA (eds) Durable resistance in crops. Plenum, New York, pp 57–80

Parlevliet JE (1985) Resistance of the non-race-specific type. In: Roelfs AP, Bushnell WR (eds) The cereal rusts, vol 2. Disease, distribution epidemiology and control. Academic Press, New York London, Orlando, pp 501–525

Parlevliet JE, Zadoks JC (1977) The integrated concept of disease resistance; a new view including horizontal and vertical resistance in plants. Euphytica 26:5–21

Payne PI, Holt LM, Johnson R, Snape JW (1986) Linkage maping of four gene loci, *Glu*-B1, *Gli*-B1, *Rg*1 and *Yr*10 on chromosome 1B of bread wheat. Genet Agrar 40:231–242

Person C (1959) Gene-for-gene relationships in host:parasite systems. Can J Bot 37:1101–1130

Person C, Sidhu G (1971) Genetics of host-parasite interrelationships. In: IAEA (ed) Mutation breeding for disease resistance. IAEA, Vienna, pp 31–38

Person C, Samborski DJ, Rohringer R (1962) The gene-for-gene concept. Nature (London) 194:561–562

Person C, Fleming R, Cargeeg L, Christ B (1983) Present knowledge and theories concerning durable resistance. In: Lamberti F, Waller JM, Van der Graaff NA (eds) Durable resistance in crops. Plenum, New York, pp 27–40

Peterson RF, Campbell AB, Hannah AE (1948) A diagrammatic scale for estimating rust intensity on leaves and stems of cereals. Can J Res (C) 26:496–500

Pope WK (1968) Interaction of minor genes for resistance to stripe rust in wheat. In: Finlay KW, Shepherd KW (eds) 3rd Int Wheat genetics Symp, Canberra, pp 251–257

Pridham JT (1939) A successful cross between *Triticum vulgare* and *Triticum timopheevii*. J Aust Inst Agric Sci 5:160–161

Priestley RH (1978) Detection of increased virulence in populations of wheat yellow rust. In: Scott PR, Bainbridge A (eds) Plant disease epidemiology. Blackwell, Oxford, pp 63–70

Priestley RH (1981) Choice and deployment of resistant cultivars for cereal disease control. In: Jenkyn JF, Plumb RT (eds) Strategies for the control of cereal disease. Blackwell, Oxford, pp 65–72

Rajaram S, Luig NH (1972) The genetic basis for low coefficient of infection to stem rust in common wheat. Euphytica 21:363–376

Rajaram S, Torres E (1986) An integrated approach to breeding for disease resistance: the CIMMYT wheat experience. In: Smith EL (ed) Genetic improvement in yield of wheat. Crop Sci Soc Am, Madison, Spec Publ 13

Rajaram S, Singh RP, Torres E (1988) Current CIMMYT approaches in breeding wheat for rust resistance. In: Simmonds NW, Rajaram S (eds) Breeding strategies for resistance to the rusts of wheat. CIMMYT, Mexico, DF, pp 101–118

Rappaport J (1954) In vitro culture of plant embryos and factors controlling their growth. Bot Rev 20:201–225

Reddy MSS, Rao MV (1979) Resistance genes and their deployment for control of leaf rust of wheat. Indian J Genet Plant Breed 39:359–365

Riley R (1958) Chromosome pairing and haploids in wheat. Proc 10th Int Congr Genetics, Montreal 2:234–235

Riley R (1965) Cytogenetics and the evolution of wheat. In: Hutchinson J (ed) Essays on crop plant evolution. Cambridge Univ Press, London, pp 103–122

Riley R (1966) Cytogenetics and wheat breeding. Contemp Agric 11–12:107–117

Riley R, Chapman V (1958) Genetic control of the cytologically diploid behavior of hexaploid wheat. Nature (London) 182:713–715

Riley R, Chapman V (1967) The inheritance in wheat of crossability with rye. Genet Res 9:259–267

Riley R, Chapman V, Johnson R (1968a) Introduction of yellow rust resistance of *Aegilops comosa* into wheat by genetically induced homoeologous recombination. Nature (London) 217:383–384

Riley R, Chapman V, Johnson R (1968b) The incorporation of alien disease resistance in wheat by genetic interference with the regulation of meiotic chromosome synapsis. Genet Res 12:199–219
Rizvi SSA, Statler GD (1982) Probable genotypes of hard red spring wheats for resistance to *Puccinia recondita* f.sp. *tritici*. Crop Sci 22:1167–1170
Robbelen G, Sharp EL (1978) Mode of inheritance, interaction and application of genes conditioning resistance to yellow rust. Fortschritte der Pflanzenzüctung, Heft 9. Parey, Berlin Hamburg
Robinson RA (1980) New concepts in breeding for disease resistance. Annu Rev Phytopathol 18:189–210
Roelfs AP (1978) Estimated losses caused by rust in small grain cereals in the United States – 1918–76. USDA Misc Publ 1363, 85 pp
Roelfs AP (1982) Effects of barberry eradication on stem rust in the United States. Plant Dis 66:177–181
Roelfs AP (1984) Race specificity and methods of study. In: Roelfs AP, Bushnell WR (eds) The cereal rusts, vol 1. Origin, specificity, structure and physiology. Academic Press, New York London Orlando, pp 131–164
Roelfs AP (1985) Wheat and rye stem rust. In: Roelfs AP, Bushnell WR (eds) The cereal rusts, vol 2. Diseases, distribution, epidemiology and control. Academic Press, New York London Orlando, pp 3–37
Roelfs AP, Groth JV (1980) A comparison of virulence phenotypes in wheat stem rust populations reproducing sexually and asexually. Phytopathology 70:855–862
Roelfs AP, Martens JW (1984) The virulence associations in *Puccinia graminis* f.sp. *tritici* in North America. Phytopathology 74:1022
Roelfs AP, McVey DV (1976) Races of *Puccinia graminis* f.sp. *tritici* in the U.S.A. during 1975. Plant Dis Rep 60:656–660
Roelfs AP, McVey DV (1979) Low infection types produced by *Puccinia graminis* f.sp. *tritici* and wheat lines with designated genes for resistance. Phytopathology 69:722–730
Roelfs AP, Long DL, Casper DH, McVey DV (1977) Races of *Puccinia graminis* f.sp. *tritici* in the U.S.A. during 1976. Plant Dis Rep 61:987–991
Roelfs AP, Baker FD, McVey DV (1982a) An interactive computer-based system for comparing cultures of *Puccinia graminis* and postularing *Sr* genotypes in wheat. Phytopathology 72:597–600
Roelfs AP, Long DL, Casper DH (1982b) Races of *Puccinia graminis* f.sp. *tritici* in the United States and Mexico in 1980. Plant Dis 66:205–207
Roelfs AP, Casper DH, Long DL (1984) Races of *Puccinia graminis* in the United States and Mexico during 1983. Plant Dis 68:902–905
Rowell JB (1981) The relationship between slow rusting and a specific resistance gene for wheat stem rust. Phytopathology 71:1184–1186
Rowell JB (1984) Controlled infection by *Puccinia graminis* f.sp. *tritici* under artificial conditions. In: Bushnell WR, Roelfs AP (eds) The cereal rusts, vol 1. Origins, specificity, structure and physiology. Academic Press, New York London Orlando, pp 291–332
Rowell JB, Hayden EB (1956) Mineral oils as carriers of uredospores of the wheat stem rust fungus for inoculating field grown wheat. Phytopathology 46:267–268
Rowell JB, Roelfs AP (1971) Evidence for an unrecognized source of overwintering wheat stem rust in the United States. Plant Dis Rep 55:990–992
Rowland GG, Kerber ER (1974) Telocentric mapping in hexaploid wheat of genes for leaf rust resistance and other characters derived from *Aegilops squarrosa*. Can J Genet Cytol 16:137–144
Saari EE, Prescott JM (1985) World distribution in relation to economic losses. In: Roelfs AP, Bushnell WR (eds) The cereal rusts, vol 2. Diseases, distribution, epidemiology and control. Academic Press, New York London Orlando, pp 259–298
Sakamura T (1918) Kurze Mitteilung über die Chromosomenzahlen und die Verwandtschaftsverhältnisse der *Triticum*-Arten. Bot Mag (Tokyo) 32:151–154
Salazar GH, Joppa LR (1981) Use of substitution-monosomics to determine the chromosome location of genes conditioning stem rust resistance in Langdon durum. Crop Sci 21:681–685
Samborski DJ (1963) A mutation in *Puccinia recondita* Rob. en Desm. f.sp. *tritici* to virulence on Transfer, Chinese Spring × *Aegilops umbellulata* Zhuk. Can J Bot 41:475–479
Samborski DJ (1980) Occurrence and virulence of *Puccinia recondita* in Canada in 1979. Can J Plant Pathol 2:246–248
Samborski DJ (1985) Wheat leaf rust. In: Roelfs AP, Bushnell WR (eds) The cereal rusts, vol 2. Diseases, distribution, epidemiology, and control. Academic Press, New York London Orlando, pp 39–59

Samborski DJ, Dyck PL (1968) Inheritance of virulence in wheat leaf rust on the standard differential wheat varieties. Can J Genet Cytol 10:24–32

Samborski DJ, Dyck PL (1976) Inheritance of virulence in *Puccinia recondita* on six backcross lines of wheat with single genes for resistance to leaf rust. Can J Bot 54:1666–1671

Samborski DJ, Dyck PL (1982) Enhancement of resistance to *Puccinia recondita* by interactions of resistance genes in wheat. Can J Plant Pathol 4:152–156

Sax K (1922) Sterility in wheat hybrids. II. Chromosome behavior in partially sterile hybrids. Genetics 7:513–552

Schafer JF, Caldwell RM, Patterson FL, Compton LE (1963) Wheat leaf rust resistance combinations. Phytopathology 53:569–573

Schafer JF, Roelfs AP, Bushnell WR (1984) Contributions of early scientists to knowledge of cereal rusts. In: Bushnell WR, Roelfs AP (eds) The cereal rusts, vol 1. Origins, specificity, structure and physiology. Academic Press, New York London Orlando, pp 3–38

Sears ER (1952) Homoelogous chromosomes in *Triticum aestivum*. Genetics 37:624

Sears ER (1953) Nullisomic analysis in common wheat. Am Nat 87:245–252

Sears ER (1954) The aneuploids of common wheat. Missouri Agric Exp Stn Res Bull 572, 58 pp

Sears ER (1956) The transfer of leaf rust resistance from *Aegilops umbellulata* to wheat. Brookhaven Symp Biol 9:1–21

Sears ER (1959) The aneuploids of common wheat. In: Jenkins BC (ed) Proc 1st Int Wheat genetics Symp, Winnipeg, pp 221–228

Sears ER (1962) The use of telocentrics in linkage mapping. Genetics 47:983

Sears ER (1966a) Chromosome mapping with the aid of telocentrics. In: Mac Key J (ed) Proc 2nd Int Wheat genet Symp, Lund. Hereditas Suppl 2:370–381

Sears ER (1966b) Nullisomic-tetrasomic combinations in hexaploid wheat. In: Riley R, Lewis KR (eds) Chromosome manipulations and plant genetics. Oliver & Boyd, Edinburgh

Sears ER (1972a) Chromosome engineering in wheat. Stadler Genet Symp 4:23–28

Sears ER (1972b) Reduced proximal crossing over in telocentric chromosomes. Genet Iber 24:233–239

Sears ER (1973) *Agropyron* wheat transfers induced by homoeologous pairing. In: Sears ER, Sears LMS (eds) 4th Int Wheat genetics Symp, Columbia, pp 191–199

Sears ER (1976) Genetic control of chromosome pairing in wheat. Annu Rev Genet 10:31–51

Sears ER (1977) An induced mutant with homoeologous pairing in common wheat. Can J Genet Cytol 19:585–593

Sears ER (1981) Transfer of alien genetic material to wheat. In: Evans LT, Peacock WJ (eds) Wheat science — today and tomorrow. Univ Press, Cambridge, pp 75–89

Sears ER (1982) A wheat mutation conditioning an intermediate level of homoeologous chromosome pairing. Can J Genet Cytol 24:715–719

Sears ER (1984) Mutations in wheat that raise the level of meiotic chromosome pairing. In: Gustafson JP (ed) Gene manipulation in plant improvement. Plenum, New York, pp 295–300

Sears ER, Okamoto M (1958) Intergenomic relationships in hexaploid wheat. Proc 10th Int Congr Genetics, Montreal 2:258–259

Sears ER, Sears LMS (1979) The telocentric chromosomes of common wheat. In: Ramanujam S (ed) Proc 5th Int Wheat genetics Symp, New Delhi, pp 389–407

Sears ER, Loegering WQ, Rodenhiser HA (1957) Identification of chromosomes carrying genes for stem rust resistance in four varieties of wheat. Agron J 49:208–212

Sharma HC, Gill BS (1983) Current status of wide hybridization in wheat. Euphytica 32:17–31

Sharma HC, Gill BS (1986) The use of *ph1* gene in direct genetic transfer and search for *Ph*-like genes in polyploid *Aegilops* species. Z Planzenzücht 96:1–7

Sharma D, Knott DR (1966) The transfer of leaf-rust resistance from *Agropyron* to *Triticum* by irradiation. Can J Genet Cytol 8:137–143

Sharp EL (1979) Male sterile facilitated recurrent selection populations for developing broad-based resistance by major and minor effect genes. Phytopathology 69:1045

Sharp EL, Smith FG (1957) Further study of the preservation of *Puccinia* uredospores. Phytopathology 47:423–429

Sharp EL, Volin RB (1970) Additive genes in wheat conditioning resistance to stripe rust. Phytopathology 60:1146–1147

Sharp EL, Schmitt CG, Staley GM, Kingsolver CH (1958) Some critical factors involved in establishment of *Puccinia graminis* var. *tritici*. Phytopathology 48:469–474

Sheen SJ, Snyder LA (1964) Studies on the inheritance of resistance to six stem rust cultures using chromosome substitution lines of a Marquis wheat selection. Can J Genet Cytol 6:74–82

Shepherd KW, Islam AKMR (1981) Wheat: barley hybrids – the first eighty years. In: Evans LT, Peacock WJ (eds) Wheat science – today and tomorrow. University Press, Cambridge, pp 107–128

Singh RP (1984) Genetics of rust resistance in wheat. PhD Thesis, Univ Sydney

Singh RP, McIntosh RA (1984a) Complementary genes for reaction to *Puccinia recondita tritici* in *Triticum aestivum.* I. Genetic and linkage studies. Can J Genet Cytol 26:723–735

Singh RP, McIntosh RA (1984b) Complementary genes for reaction to *Puccinia recondita tritici.* II. Cytogenetic studies. Can J Genet Cytol 26:736–742

Singh RP, McIntosh RA (1986) Cytogenetical studies in wheat. XIV. *Sr8b* for resistance to *Puccinia graminis tritici.* Can J Genet Cytol 28:189–197

Skovmand B, Roelfs AP, Wilcoxson RD (1978a) The relationship between slow-rusting and some genes specific for stem rust resistance in wheat. Phytopathology 68:491–499

Skovmand B, Wilcoxson RD, Shearer BL, Stucker RE (1978b) Inheritance of slow rusting to stem rust in wheat. Euphytica 27:95–107

Smith EL, Schlehuber AM, Young HC, Jr. Edwards LH (1968) Registration of Agent wheat. Crop Sci 8:511–512

Snape JW, Law CN (1980) The detection of homologous chromosome variation in wheat using backcross reciprocal monosomic lines. Heredity 45:187–200

Snape JW, Parker BB, Gale MD (1983) Use of the backcross reciprocal monosomic method for evaluating chromosomal variation for quantitative characters. In: Sakamoto S (ed) Proc 6th Int Wheat genetics Symp, Kyoto, pp 367–373

Soliman AS, Heyne EG, Johnston CO (1964) Genetic analysis for leaf rust resistance in the eight differential varieties of wheat. Crop Sci 4:246–248

Stakman EC (1914) A study in cereal rusts: Physiological races. Minn Agric Exp Stn Bull 138

Stakman EC, Levine MN (1922) The determination of biologic forms of *Puccinia graminis* on *Triticum* spp. Univ Minn Agric Exp Stn Tech Bull 8: 10 pp

Stakman EC, Piemeisel FJ (1917) Biologic forms of *Puccinia graminis* on cereals and grasses. J Agric Res 10:429–495

Stakman EC, Piemeisel FJ, Levine MN (1918) Plasticity of biologic forms on *Puccinia graminis.* J Agric Res 15:221–250

Stakman EC, Levine MN, Leach JG (1919) New biologic forms of *Puccinia graminis.* J. Agric Res 16:103–105

Stakman EC, Levine MN, Griffee F (1925) Webster, a common wheat resistant to black stem rust. Phytopathology 15:691–698

Stakman EC, Kempton FE, Hutton D (1927) The common barberry and black stem rust. US Dep Agric Farmers' Bull 1544: 28 pp

Stakman EC, Stewart DM, Loegering WQ (1962) Identification of physiological races of *Puccinia graminis* var. *tritici.* US Dep Agric ARS E617: 53 pp

Staples RC, Macko V (1984) Germination of urediospores and differentiation of infection structures. In: Bushnell WR, Roelfs AP (eds) The cereal rusts, vol 1. Origins, specificity, structure, and physiology. Academic Press, New York London Orlando, pp 255–289

Statler GD (1977) Inheritance of virulence of culture 73–47 *Puccinia recondita.* Phytopathology 67:906–908

Statler GD (1979) Inheritance of pathogenicity of culture 70–1, race 1, of *Puccinia recondita tritici.* Phytopathology 69:661–663

Statler GD (1984) Probable genes for leaf rust resistance in several hard red spring wheats. Crop Sci 24:883–886

Statler GD, Jones DA (1981) Inheritance of virulence and uredial color and size in *Puccinia recondita tritici.* Phytopathology 71:652–655

Steel RGD, Torrie JH (1980) Principles and procedures of statistics, 2nd edn. McGraw-Hill, New York

Stubbs RW (1985) Stripe rust. In: Roelfs AP, Bushnell WR (eds) The cereal rusts, vol 2. Diseases, distribution, epidemiology, and control. Academic Press, New York London Orlando, pp 61–101

Taylor EC (1976) The production and behavior of somatic recombinants in *Puccinia striiformis.* In: Brönnimann A (ed) Proc 4th European and Mediterranean cereal rust Conf, Intertaken, pp 36–38

Tee TS, Qualset CO (1975) Bulk methods in wheat breeding: comparison of single seed descent and random bulk methods. Euphytica 24:393–405

Tervet IW, Rawson AJ, Cherry E, Saxon RB (1951) A method for the collection of microscopic particles. Phytopathology 41:282–285

The TT (1973) Chromosome location of genes conditioning stem rust resistance transferred from diploid to hexaploid wheat. Nature New Biol 241:256

Thiebaut J, Kasha KJ (1978) Modification of the colchicine technique for chromosome doubling of barley haploids. Can J Genet Cytol 20:513–521

Thomas JB, Kaltsikes PJ, Anderson RG (1981) Relation between wheat-rye crossability and seed set of common wheat after pollination with other species in the Hordeae. Euphytica 30:121–127

Tsunewaki K (ed) (1980) Genetic diversity of the cytoplasm in *Triticum* and *Aegilops* Jpn Soc Promo Sci. Tokyo

Tiyawalee D, Frey KJ (1970) Mass selection for crown rust resistance in an oat population. Iowa State J Sci 45:217–231

Unrau J (1958) Genetic analysis of wheat chromosomes. Can J Plant Sci 38:415–418

Unrau J, Person C, Kuspira J (1956) Chromosome substitution in hexaploid wheat. Can J Bot 34:629–640

Vakili NG, Galdwell RG (1957) Recombination of spore color and pathogenicity between uredial clones of *Puccinia recondita* f.sp. *tritici.* Phytopathology 47:536 (Abstr)

Valkoun J, Bartos P (1974) Somatic chromosome number in *Puccinia recondita.* Trans Br Mycol Soc 63:187–189

Vallega V (1979) Search for useful characters in diploid *Triticum* spp. In: Ramanujam S (ed) Proc 5th Intern Wheat Genet Symp, New Delhi, pp 156–162

Vanderplank JE (1963) Plant diseases: epidemics and control. Academic Press, New York London

Vanderplank JE (1968) Disease resistance in plants. Academic Press, New York London

Vanderplank JE (1975) Principles of plant infection. Academic Press, New York San Francisco London

Vanderplank JE (1978) Genetic and molecular basis of plant pathogenesis. Springer, Berlin Heidelberg New York

Vanderplank JE (1982) Host-pathogen interactions in plant disease. Academic Press, New York London

Vanderplank JE (1983) Durable resistance in crops: should the concept of physiological races die? In: Lamberti F, Waller J, Van der Graaff NA (eds) Durable resistance in crops. Plenum, New York, pp 41–44

Vanderplank JE (1984) Disease resistance in plants, 2nd edn. Academic Press, New York London Orlando

Vanderplank JE (1985) Virulence structure in *Puccinia graminis* f.sp. *tritici:* A reply. Phytopathology 75:109

Wall AM, Riley R, Chapman V (1971a) Wheat mutants permitting homoeologous meiotic chromosome pairing. Genet Res 18:311–328

Wall AM, Riley R, Gale MD (1971b) The position of a locus on chromosome 5B of *Triticum aestivum* affecting homoeologous chromosome pairing. Genet Res 18:329–339

Wallwork H (1982) Transgressive segregation for resistance to yellow rust in wheat. PhD Thesis, Univ Cambridge

Wallwork H, Johnson R (1984) Transgressive segregation for resistance to yellow rust in wheat. Euphytica 33:123–132

Ward HM (1903) Further observations on the brown rust of the bromes, *Puccinia dispersa* (Erikss.) and its adaptive parasitism. Ann Mycol 1:132–151

Washington WJ, Maan SS (1974) Disease reaction of wheat with alien cytoplasms. Crop Sci 14:903–905

Watson IA (1957) Further studies on the production of new races from mixtures of races of *Puccinia graminis* var. *tritici* on wheat seedlings. Phytopathology 47:510–512

Watson IA (1981) Wheat and its rust parasites in Australia. In: Evans LT, Peacock WJ (eds) Wheat science – today and tomorrow. Univ Press, Cambridge, pp 129–147

Watson IA, Baker EP (1943) Linkage of resistance to *Erysiphe graminis tritici* and *Puccinia triticina* in certain varieties of *Triticum vulgare* Vill. Proc Linn Soc NSW 68:150–152

Watson IA, Luig NH (1958) Somatic hybridization in *Puccinia graminis tritici.* Proc Linn Soc NSW 83:190–195

Watson IA, Luig NH (1961) Leaf rust on wheat in Australia: a systematic scheme for the classification of strains. Proc Linn Soc NSW 86:241–250

Watson IA, Luig NH (1963) The classification of *Puccinia graminis* var. *tritici* in relation to breeding resistant varieties. Proc Linn Soc NSW 88:235–258

Watson IA, Luig NH (1966) *Sr15* – a new gene for use in the classification of *Puccinia graminis* var. *tritici*. Euphytica 15:239–250

Watson IA, Luig NH (1968) Progressive increase in virulence on *Puccinia graminis* f.sp. *tritici*. Phytopathology 58:70–73

Wehrhahn C, Allard RW (1965) The detection and measurement of the effects of individual genes involved in the inheritance of a quantitative character in wheat. Genetics 51:109–119

Wellings CR, McIntosh RA (1982) Stripe rust – a new challenge to the wheat industry. Agric Gaz NSW 92:2–4

Wellman FL, Blaisdell DJ (1940) Differences in growth characters and pathogenicity of Fusarium wilt isolations tested on three tomato varieties. US Dep Agric Tech Bull 705

Wienhues A (1979a) Translokationslinien mit Resistenz gegen Braunrost (*Puccinia recondita*) aus *Agropyron intermedium*. Ergebnisse aus der Rückkreuzung mit Winterweizensorten. Z Pflanzenzücht 82:149–161

Wienhues A (1979b) Resistenz gegen Gelbrost (*Puccinia striiformis*) aus *Agropyron intermedium* übertragen in den Winterweizen. Z Pflanzenzücht 82:201–211

Wilcoxson RD, Skovmand B, Atif AH (1975) Evaluation of wheat cultivars for ability to retard development of stem rust. Ann Appl Biol 80:275–281

Williams E, Jr., Johnston CO (1965) Effect of certain temperatures on identification of physiologic races of *Puccinia recondita* f.sp. *tritici*. Phytopathology 55:1317–1319

Williams ND, Gough FJ, Rondon MR (1966) Interaction of pathogenicity genes in *Puccinia graminis* f.sp. *tritici* and reaction genes in *Triticum aestivum* spp. *vulgare* 'Marquis' and 'Reliance'. Crop Sci 6:245–248

Williams PG (1975) Evidence for diploidy of a monokaryotic strain of *Puccinia graminis* f.sp. *tritici*. Trans Br Mycol Soc 64:15–22

Williams PG (1984) Obligate parasitism and axenic culture. In: Bushnell WR, Roelfs AP (eds) The cereal rusts, vol 1. Origins, specificity, structure, and physiology. Academic Press, New York London Orlando, pp 399–430

Williams PG, Scott KJ, Kuhl JL (1966) Vegetative growth of *Puccinia graminis* f.sp. *tritici* in vitro. Phytopathology 56:1418–1419

Winkle WE, Kimber G (1976) Colchicine treatment of hybrids in the Triticinae. Cereal Res Commun 4:317–320

Wolfe MS (1978) Some pratical implications of the use of cereal variety mixtures. In: Scott PR. Bainbridge A (eds) Plant disease epidemiology. Blackwell, Oxford, pp 201–207

Wolfe MS (1985) The current status and prospects of multiline cultivars and variety mixtures for disease resistance. Annu Rev Phytopathol 23:251–273

Wolfe MS, Knott DR (1982) Populations of plant pathogens: some constraints on analysis of variation in pathogenicity. Plant Pathol 31:79–90

Wolfe MS, Schwarzbach E (1975) The use of virulence analysis in cereal mildews. Phytopathol Z 82:297–307

Wolfe MS, Barrett JA, Shattock RC, Shaw DS, Whitbread R (1976) Phenotype-phenotype analysis: field applications of the gene-for-gene hypothesis in host-pathogen relations. Ann Appl Biol 82:369–374

Wolfe MS, Minchin PN, Slater SE (1981a) Powdery mildew of barley. Plant Breed Inst Annu Rep 1980:88–92

Wolfe MS, Barrett JA, Jenkins JEE (1981b) The use of cultivar mixtures for disease control. In: Jenkyn JF, Plumb RT (eds) Strategies for the control of cereal disease. Blackwell, Oxford, pp 73–80

Worland AJ, Law CN (1986) Genetic analysis of chromosome 2D of wheat. I: The location of genes affecting height, day-length insensitivity, hybrid dwarfism and yellow rust resistance. Z Pflanzenzücht 96:331–345

Worland AJ, Law CN (1987) Seedling resistance to yellow rust. Plant Breed Inst Annu Rep 1986:74–75

Worland AJ, Gale MD, Law CN (1987) Wheat genetics. In: Lupton FGH (ed) Wheat breeding: its scientific basis. Chapman & Hall, London New York, pp 129–171

Wright RG (1976) Variation in *Puccinia striiformis*. In: Brönnimann A (ed) Proc 4th European and Mediterranean cereal rusts Conf, Interlaken, pp 42–44

Wright RG, Lennard JH (1978) Mitosis in *Puccinia striiformis* 1. Light microscopy. Trans Br Mycol Soc 70:91–98

Wright RG, Lennard JH (1980) Origin of a new race of *Puccinia striiformis*. Trans Br Mycol Soc 74:283–287

Young HC, Jr., Browder LE (1965) The North American 1965 set of supplemental differential wheat varieties for identification of races of *Puccinia recondita tritici*. Plant Dis Rep 49:308–311

Zadoks JC (1961) Yellow rust on wheat studies in epidemiology and physiologic specialization. Tijdschr Plantenziek 67:69–256

Zeller FJ (1973) 1B/1R wheat-rye substitutions and translocations. In: Sears ER, Sears LMS (eds) Proc 4th Int Wheat genetics Symp, Columbia, pp 209–221

Zeller FJ (1981) Identification of a 4A/7R and a 7B/4R wheat-rye chromosome translocation. Theor Appl Genet 59:33–37

Subject Index